P9-CQI-646

# The Ocean World of Jacques Cousteau

# The Art of Motion

# The Ocean World of Jacques Cousteau

## Volume 5

## The Art of Motion

*Flipping their tails from side to side and using their fins for stability, these soldierfish cavort among coral growths. By one means or another, at one speed or another, most creatures of the sea enjoy a three-dimensional freedom of movement.*

The Danbury Press
A Division of Grolier Enterprises Inc.

Publisher: Robert B. Clarke

Production Supervision: William Frampton

---

Published by Harry N. Abrams, Inc.

Published exclusively in Canada by
Prentice-Hall of Canada, Ltd.

Revised edition—1975



---

Project Director: Peter V. Ritner

Managing Editor: Steven Schepp
Assistant Managing Editor: Ruth Dugan
Senior Editors: Donald Dreves
Richard Vahan
Assistant Editors: Jill Fairchild
Sherry Knox

Creative Director and Designer: Milton Charles

Assistant to Creative Director: Gail Ash
Illustrations Editor: Howard Koslow

Production Manager: Bernard Kass

Science Consultant: Richard C. Murphy

Printed in the United States of America

234567899876

LIBRARY OF CONGRESS CATALOGING
IN PUBLICATION DATA

Cousteau, Jacques Yves.
The art of motion.

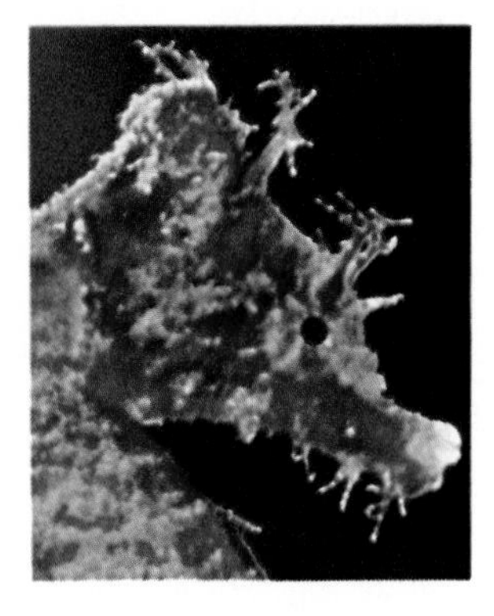

# Contents

One of the key factors to the behavior of animals in general is their talent for self-propulsion. In the sea where it originated, LIFE MOVES, and though we may not know exactly how the art of motion entered into our evolutionary heritage, there is a strong probability that propulsion and digestion both originated from the most simple rhythmic contractions. 8

In the sea all motion must be performed through THE DENSE MEDIUM (Chapter I). But the very density of water also has advantages. It buoys up immersed bodies, so that much of the energy land animals must devote to combat the forces of gravity can be turned to other more useful purposes. 10

To improve progression through water, the various formulas that have been developed through the ages have concurrently achieved two major goals in FORM AND DESIGN (Chapter II): (1) that body shapes should slip through water creating as little resistance as possible and (2) that efficient power plants and power transducers should be compatible with the above body styling. Modern hydrodynamic engineers, trying to reach the same basic results for man's ships, have either from independent research or from imitations of nature designed hulls that resemble various marine creatures. 18

The swimming performances of some undersea creatures are so remarkable that only recently have we achieved a physical understanding of them. In the TRAVELING WAVE (Chapter III), fin movements are coordinated with body undulations to create a wave that increases in amplitude and decreases in wavelength as it approaches the tail. 30

Body shape is the passive part of the story. The power plant is the source of propulsion. THE ROLE OF FINS (Chapter IV) is to translate energy into thrust, but also to lift, stabilize, stop, hover, and turn. 38

Some animals have learned to huff and puff their way through the water, either using their mouths or better yet some specialized structure like a rocket nozzle. These are THE JET SET (Chapter V), employing virtually the same principle of action and reaction used on man's most advanced aircraft. Octopus and squid are particularly successful jetters. 54

Then there are the proletarians of the sea, the WALKERS AND CRAWLERS (Chapter VI). The awkward progress made by the mudskipper on his pelvic fins may be somewhat

similar to the first steps of the remote ancestors of land vertebrates—and his fins may reflect the origin of our limbs. 62

Some creatures have chosen OTHER WAYS TO GO (Chapter VII): the waggling iguana, the surprisingly graceful walrus, the webbed-footed birds, the flying mantas, the paddling turtles—and even those coelenterates which create sails for themselves in order to be blown about the sea. Men have learned from nature, and have access from their undersea crafts to animal technology—sometimes with more power, but always with a lot less maneuverability. 76

An attractive way of life is that of HITCHHIKERS (Chapter VIII). As is also the case with human hitchhikers, some of them underwater are dangerous—parasites that feed on the blood and flesh of their hosts, in extreme cases killing them. Others, like the remora, just ride along doing no injury to and conferring little benefit on the larger animals. 88

When the flyingfish launches himself into the air, it is no game: he is escaping from some predator unseen beneath the surface, literally GETTING THE HULL OUT OF WATER (Chapter IX). This has the obvious advantage of reducing drastically the hydrodynamic drag. Many fish, like the mullet, leap for survival; out of the water they are temporarily invisible to whatever is chasing them. The great mammals, whales as well as sea lions and seals, leap for play, or to "spy hop," or to grab a breath of air. Man imitates the flyingfish with his Hovercrafts and hydrofoils. 94

Modern man, of course, is always aiming TOWARD HIGHER SPEEDS (Chapter X). And most of his best clues have come from the sea: turbulence-damping shapes and surfaces, combination of modes of propulsion, propellers acting as rotating fins, paddles resembling flippers, airborne ships, and jet propulsion. 112

The art of motion also demonstrates that MOVEMENT SHAPES A MODE OF LIFE (Chapter XI). Animals which hunt their food in the open sea tend to collect the equipment and hydrodynamic form needed for great speed. Animals that crawl along the bottom are fitted with clumsier, but more heavily armored forms. 130

The need for MOVING ELSEWHERE has produced, over more than a billion years, a multitude of near-perfect designs, either fitted to or generating all conceivable behaviors. 142

# Introduction: Life Moves

One of the clichés of the horror filmmaker is the mobile vegetable, especially if it is carnivorous! Most of us are willing to accept—not perhaps as casually as we ought to—the contortions of serpents, the eight-legged hops of tarantulas, the multilimbed progress of centipedes, the sinister creep of the brittle star. But that a carrot should painfully pull itself up out of the ground and wriggle towards us: that is terrifying. It was, after all, the ambulatory wood of Birnam which persuaded Macbeth that all was lost. Plants should stay put, our instincts tell us. Movement is for animals!

Motion is such a critical component of animalness that it deserves, and gets, a huge body of special research. And this study has yielded some particularly enlightening results in the undersea world, because in the ocean the general challenges of propulsion are complicated by the density of the medium through which the motion must be made. In the sea we find not one but three "tops of the line," three elaborate near-perfect animal machines—each employing a different mode of propulsion. First, the ultimate jet-propelled creature: the giant squid. Second, the ultimate streamlined cruiser: the tuna. Third, the ultimate high-efficiency thermodynamic machine: the dolphin.

Where did these three lines of development originate? How did propulsion come to be a feature of the animal world, how does it fit into the evolutionary picture? It is not possible to answer these questions with certainty. But we may find clues by imagining the life-style of an extremely primitive one-celled creature, looking much like our familiar contemporary, the amoeba. This infinitesimal protoanimal must eat. As he wraps his protoplasmic arms around an even smaller prey, or after digestion rejects the unusable excrement, tiny spasms ripple over his cellular membrane imparting an impulse to the little body. He begins to drift through the watery wastes. Soon he finds himself in another part of the sea with fresh food supplies. Again he feeds, again his cell walls convulse, again he forges on to new regions.

In other words, the act of nutrition itself may imply propulsion in animals. As the millennia and the millions of years pass by, those amoebalike animals which move sluggishly, and therefore encounter less food, tend to die out. Those that move faster find more extensive grazing and tend to survive. Later another theme joins the evolutionary tale. Those animals which learn to *sense* a food particle, and to move *purposively* toward it, not only survive but set foot on the path leading to nervous systems and the manifold differentiated forms of animal life we find in the fossil record and around us on earth today. Those animals that can move but that develop little if any motivated control of their movements—which means that they must count on chance to carry food to them—tend to die out or to invest in alternate strategies of species survival: i.e., sheer fecundity.

All really efficient systems of propulsion are found in the ocean. In open water the two main mechanical ones are the jet system and the body-fin-undulation system—represented by the squid and the tuna. In addition, countless minor systems have been spun off to cope with other styles and niches of life: crawling, walking, jumping, burrowing, etc., etc. In the case of the third "top of the line" animal—the dolphin—superb propulsion equipment is supplemented by related systems that we will discuss in detail later: a complex skin structure that damps the turbulence the dolphin's velocity creates; and the mammalian warm-blood physiology that enables the dolphin to function with exceptionally high efficiency across a wide range of temperatures.

It is fascinating to trace the parallels between the naturally evolving modes of propulsion in the sea and man's artificial ones. Man swims, but with ridiculous clumsiness as compared to any fish. Yet the skin diver's flippers are surrogate fins. The wet suit is the analogue of the dolphin's and sea lion's insulating skin and blubber layer. The jet principle, of course, man reserves mainly for travel through the air. Our engineers have not achieved the skills necessary for constructing a vehicle which moves by undulation, any more than they have built a machine which truly flies as does a bird. Instead, to propel most surface ships and submarines, we have coupled a rotating fin—the propeller—to highly inefficient forms of power like the reciprocating steam-engine or the internal combustion engine. Modern nuclear plants are also driving ships with propellers. Jet units are only used on sophisticated speedboats and Hovercraft. In terms of shapes our borrowings from the sea have come easier. The same hydrodynamical principles that over the ages have polished the streamlined forms of the great fish are applied in designing hulls, rudders, keels, or planing surfaces.

One of these days man will voyage from solar system to solar system in vessels powered by gigantic sails that catch and direct the ions flowing out into space from all the active stars. In theory ionic propulsion is capable of speeds very close to the speed of light. Even this has its underwater inspiration. One of Albert Einstein's earliest scientific papers concerned Brownian motion—that random, never-ceasing dance of microscopic particles in a fluid as they are struck by the molecules of the fluid. Here again, at the outset of the career which transformed our world, the sea provided the clue.

Jacques-Yves Cousteau

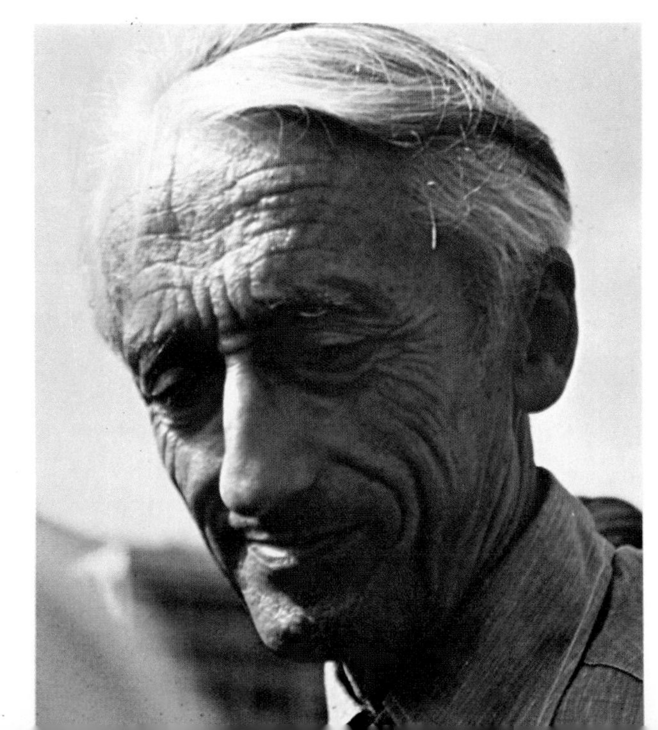

# Chapter I. The Dense Medium

Until we engage in some strenuous physical activity, most of us never think about how much effort and energy we use just standing around. Land-dwelling animals, like ourselves, constantly fight gravity—the force of the earth that untiringly pulls everything toward its center. When we climb a flight of stairs, we lift our entire weight with each step. When we walk across a room, our legs and feet must lift our weight and push it off the floor. Even standing still doesn't free us from gravity's bond — our legs must continue supporting our weight and thousands of small muscles around our veins must push the blood up toward the heart. While we must constantly exert ourselves to remain upright, we have no difficulty moving at low speeds through the air surrounding us. Seawater, on the contrary, is 800 times more dense than air and greatly hinders movement through it, as you know if you have ever tried to walk through waist-deep water.

**"In water a buoyant force pushes objects toward the surface, counteracting the force of gravity."**

The water in the oceans is the same as that in lakes and streams, but ocean water has a greater quantity of materials dissolved in it. Most significant of these are the salts. Salt increases the density of water, so ocean dwellers have a slightly more dense medium than their freshwater counterparts.

Viscosity is the internal frictional resistance of a substance, or a measure of the attraction of its molecules to one another which creates a tug on bodies attempting to move through the substance. Try drawing a spoon through a jar of honey and then through a container of water. The spoon passes through the water more easily than through the honey, because the honey is more viscous than water. But to make a fair comparison of relative viscosities and their influence on the movement of animals we should compare the viscosities of water and air, the two mediums that can sustain life.

Ocean water has a greater quantity of salts and materials dissolved in it than in lakes and streams. Salt increases the density of water, so ocean dwellers live in a slightly more dense medium than their freshwater counterparts.

In spite of the problems involved in moving through a dense and viscous medium, marine animals do enjoy a certain advantage over land dwellers. In water a buoyant force pushes objects toward the surface, counteracting the force of gravity. Because of this buoyancy, in water the specific weight (density) of a submerged object is reduced: a piece of wood will float; a squid will move around practically weightless; a three-pound aluminum ingot will weigh only two pounds; a seven-pound pig iron, only six pounds.

***Supporting world of whales.*** *The majestic California gray whale parades up and down the west coast of North America on an annual 5000-mile march. Smaller by about half than its enormous cousin, the blue whale which is the largest animal ever to inhabit the earth, the gray still weighs in at an impressive 40 tons and measures up to 50 feet in length. Several million years ago the ancestors of the whales adopted water as their living space and grew to mammoth size. In water they are freed of the force of gravity; the buoyancy of water supports their massive bodies. Out of water whales must die. When they are stranded on a beach, their lungs collapse under the great weight of their bodies and they suffocate.*

***Logs*** *float because the volume of water they displace weighs more than they do.*

## Buoyancy

In the third century B.C. Archimedes was the leading scientist at court of Hiero II of Syracuse, a Greek city in Sicily. Legend holds that the king was in the market for a new crown, one made of pure gold. A metalworker offered one to the king for an extraordinarily low price. The king suspected that this crown was not made entirely of gold and thought that silver may have been substituted. But, having no way to confirm his suspicions, he called in Archimedes to prove his theory.

Archimedes took samples of pure silver and gold, which were equal to the crown's weight. He immersed each in a water-filled container and measured the amount of water that overflowed. He discovered that the volume of water displaced by the gold sample was less than that displaced by the silver. When the crown was immersed, the volume of water that spilled from the container was less than the amount displaced by the silver, but more than that displaced by the gold. Archimedes concluded that the crown was a combination of gold and silver, and the attempted deception was uncovered.

> "Archimedes's principle: An object immersed in a liquid is buoyed up by a force equal to the weight of the volume of liquid it displaces."

***Heavier than water.*** *After centuries on the bottom of the sea, this heavy iron cannon is encrusted with marine life and is corroding away. In the sea it weighs less than it does on land because of the buoyant force of the water. The difference between the cannon's weight on land and in the sea is the weight of the water it displaces, in this case one-seventh of its weight in air.*

Archimedes's experiment demonstrated for the first time the buoyant influence of water, and the principle underlying this phenomenon was named after him. Archimedes's principle states that an object immersed in a liquid is buoyed up by a force equal to the weight of the volume of liquid it displaces. An object floats when the buoyant force exceeds its weight. It sinks when the displaced volume of liquid weighs less than it does. If an object displaces a volume of liquid that weighs exactly the same as it does, it is said to be neutrally buoyant; it neither sinks nor floats. Such neutrally buoyant objects are currently built in plastic to drift at specific depths and are used to measure deep-sea currents.

***Two gulls*** *soar in over a beach to join others already standing near the water's edge. When coming in for landings, gulls twist their wings in very much the same way an airplane lowers its wing flaps.*

## To Defy Gravity

The effortlessly soaring gull has not escaped gravity but has temporarily lifted itself into a supporting airflow. It had to beat its wings with great force and rapidity to reach the invisible cushions of air on which it rides. If the updrafts fail, the bird must again flap its wings to stay aloft, or it will sink to earth. Although its anatomy is almost perfectly suited for aerial life, a bird must expend tremendous amounts of energy to become airborne. A tiny bird burns at least one percent of its body weight for each hour of flight, and an active small bird must eat almost its body's weight of food each day. It is costly to

***These cardinalfish*** *are found hovering above the bottom on coral reefs. With the water to buoy up their tiny bodies they can remain there effortlessly and consequently need very little food.*

fly in the face of earth's gravity. To reduce that cost the central temperature of a bird is high in order to increase the energy output of the muscles, its skeleton is thin and hollow, and streamlining is achieved with feathers.

Seabirds, however, have to find their food in the sea; the stormy petrel hovers close to the surface and picks its small prey from the first inches of water; boobies and terns dive-bomb for short, fairly shallow but efficient intrusions; cormorants can stay five minutes in the sea and "swim" half a mile with their wings. Heavier than air, birds are lighter than water, and must fight their way down into the sea, as they fight their way up.

*Grunts with swim bladders, like the one above, can regulate their volume at will to stay hovering without effort at whatever level in the sea they want.*

*The **angel shark** seen at right is launching itself from the bottom with powerful sweeps of its pectorals. Because it is heavier than water and has no swim bladder, this ray must make constant efforts to stay above the sea floor.*

## Swim Bladder

The ocean is a three-dimensional world for all marine creatures. Some of these are sedentary, while others constantly travel in the open sea. Rock and reef fishes need not cover great distances; most of their life is spent moving slowly about in the vicinity of their home. They are greatly helped in their command of the third dimension by a built-in gas-filled chamber, or swim bladder: without it, they would slowly but constantly sink to the bottom, because their bodies, made of flesh and bones, are slightly heavier than sea water. The swim bladder increases their volume without increasing their weight and keeps them perfectly balanced.

> "In a rapid ascent, a fish must eliminate gas from its swim bladder very quickly to avoid ballooning."

When they descend, the gas in the bladder is compressed and more gas must be quickly generated therein in order to keep them neutrally buoyant. If they rise, some of the expanding gas has to be eliminated very rapidly to avoid ballooning to the surface with their bladders protruding from their mouths.

The mechanism of instant control of the swim bladder is not yet well understood.

## No Swim Bladder

Some of the fast open-ocean swimmers, such as certain tuna and sharks, have no such buoyancy system as a swim bladder. They all slowly sink when they stop swimming. This is rather convenient for creatures that spend a substantial part of their lives resting or hiding on the bottom: rays, skates, and sharks like the angel shark above often remain motionless for hours on the sea floor. Their apparent weight, being roughly one-twentieth of what it would be in air, is enough to anchor them against the drag of currents, tides, or swells.

For these tuna and sharks, however, a constantly negative buoyancy is a dominant factor that imprisons them in a way of life: crossing oceans for weeks or months at a time at the depths where their food can be found, but with miles of water under them, they are sentenced to swim forever and probably never to sleep.

# Chapter II. Form and Design

From a general viewpoint, in order to move, an animal must generate lift to counteract gravity, and must generate thrust to counteract drag. We have seen that fish are privileged because they are buoyed up by the water's lift and thus have to make no effort to overcome gravity. On the other hand, they are handicapped because they need to develop very substantial thrust forces to penetrate swiftly the oceans.

An Apollo spacecraft hurtles through the vacuum of space, free from the forces of gravity and from drag. All the launching's rocket energy was spent to snatch the craft from the pull of the earth and from the resistance of the atmosphere. Now, in outer space, it can maintain its speed at no expense of fuel. Its appendages stick out here and there; its shape does not matter because out of the atmosphere there is no drag, no turbulence to slow down the vehicle. But in air, propulsion obeys the laws of aerodynamics; modern aircraft design has been inspired by the shapes of such birds as the swallow or the albatross. The streamlining of a supersonic jet plane is sleek, its shape being studied in every detail to reduce turbulence and offer the least possible resistance.

In water, the laws of hydrodynamics are generally similar but much more severe, because of the high density and high viscosity of liquids. Aquatic organisms experience a very strong drag, which is a function of speed, of shape, of the surface of the body, and of the nature of the water flow over it. These four factors cannot be considered separately because they interfere heavily with one another. The drag forces become very large at rapid speeds because they increase approximately as the square of the velocity. A fusiform shape, with the thickest portion one-third of the way back, like a tunafish or a torpedo, tapered at either end, separates water easily in front and allows it to converge smoothly behind. In water as in air, it is the best streamlined design; the higher the speed, the more elongate the shape must be, with a reduced cross section, but also with an increased surface to encompass the same volume of flesh. Now, when the surface of the body increases, the drag due to friction increases, and the remedy is to improve the nature of the surface, as skiers do when they wax their skis. Finally, any structures projecting from the body tend to produce turbulence, so even fins and flippers are streamlined and eventually can be folded against the body or even inside special slots on the back. Even the inside of the mouth and especially the gills are well streamlined to reduce the drag of "inner water flow."

The basic fusiform shape may be compressed either in the vertical plane or in the horizontal plane. Those marine animals that live a sedentary life don't need speed; their shapes are not governed by hydrodynamics.

***Underwater disturbances — slow movement.*** *The wake left behind by this swan has created turbulence in otherwise still water. As the swan moves, it forces water ahead, setting up wavelets on the surface and disturbances beneath the surface. These underwater disturbances (turbulence), whether made by birds or mammals on the surface or by fishes and other animals underwater, tend to slow the swimmers, clinging to their sides, hindering their progress. In front, you can see how the water tends to furrow upward as snow does in front of a snowplow, slowing forward progress. What you can't see are the eddies, swirls, vortexes of water all around the swan. These eddies, swirls, and plowed-up water have the same effect on fishes. But fishes, living as they have in water for 350 million years, have adapted to their environment. Those that need to swim swiftly have made the necessary body and shape adaptations to cleave the water with a minimum of hindering turbulence.*

## The Price of Speed

When the Atlantic dolphin, shown here, breaks the surface, he creates a wake that generates a strong drag. When totally submerged, the dolphin does better, being able to accommodate his shape to reduce turbulence. At slow cruising speeds, the surface effect does not influence noticeably the economy of the dolphin's propulsion. But at high speeds, the dolphin jumps clear out of the water to breathe and to reduce the time during which he pulls a costly wake behind him.

Two factors can reduce the speed of marine animals in water: "pressure resistance" and "friction drag." Pressure resistance is a function of streamlining and is the consequence of having to push water ahead and force it to flow around the body. In friction drag, which mainly depends on the smoothness of the skin, some water is moving with the fish against the water farther away which is not moving, thus creating turbulence.

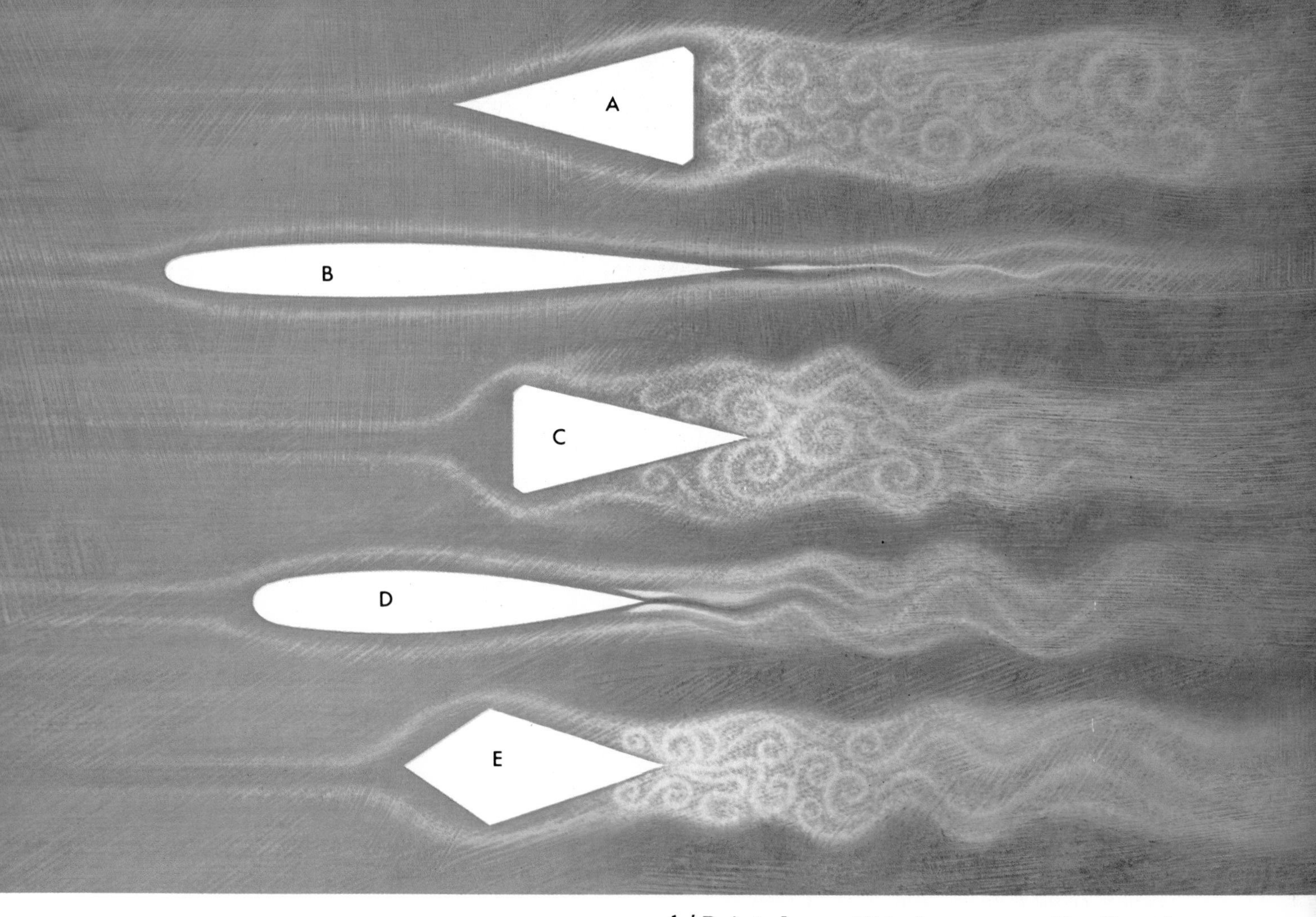


## Streamlining

Water flow can be *laminar* (smooth), *turbulent* (irregular), or *transitional*. Laminar flow creates the least drag, and turbulent flow the most. In practice, any fish or mammal or man-made hull produces transitional flow, as perfect laminar or totally turbulent flows are never encountered.

Streamlined bodies promote laminar flow and the faster an object travels through water, the more streamlined it must be.

Turbulence begins to set in when the thin layer of water immediately next to the moving body (the "boundary layer") becomes unstable. Such instability is, at least partially, inevitable in man-made rigid hulls. If the turbulence in the boundary layer can be stabilized, then laminar flow can be maintained over the entire organism. It seems that fish and marine mammals have such a stabilization mechanism: by constantly changing their shape to conform their body surfaces to the lines of flow they are able to move at speeds that could not be matched by exact, but rigid, replicas of their forms.

***A / Pointed nose.*** *This shape eases cutting through the water, but the broadness of the after portion causes heavy drag and boiling up of water just behind the body.*

***B / Long and slim.*** *This characteristic bullet shape of the fusiform body like that of a barracuda or a shark. It creates less turbulence as this drawing based on a photographic study indicates.*

***C / Blunt front.*** *This is the least efficient shape shown; it stirs water and creates the most turbulence.*

***D / Shortened and broadened.*** *This shape creates greater turbulence than does the fusiform body.*

***E / Angular on the sides.*** *This shape creates still greater turbulence.*

## Fish Shapes

Besides displaying a wide variety of sizes and colors, fish come in many shapes. Where the fish lives, how it feeds, what it eats, how fast it swims, and what its relationships to other animals are — all may be affected by its body shape. The shape of fish's bodies fall into general categories: fusiform, like the shark, barracuda, and codfish; laterally (side to side) compressed, like the angelfish, spadefish, and filefish; dorsoventrally (top to bottom) compressed, like the skate and guitarfish; attenuated, like the conger or the American eel; and a number of other shapes that might be called miscellaneous because there are only a few examples of each (these might include the strangely shaped seahorse, the triangular cowfish, and the globular porcupinefish). Whatever their shape, fish are similar in their bilateral symmetry.

Although we have given general categories of fish shapes, most species exhibit characteristics of more than one category. Few fish, for example, are precisely tubular, thus fitting the exact definition of a fusiform fish. Most that come close to this shape are usually somewhat flattened dorsoventrally or laterally. A few are long and drawn out as well as tubular and therefore combine fusiform and attenuated shapes.

The fish has assumed its present shape through many millions of years of natural selection. That is, the individuals of each species best suited for their particular environment had a better chance to survive long enough to reproduce and pass on their genetic material to their offspring, who then did the same. Those less suited either moved to more suitable environments or died before reproducing and passing their genes to offspring.

**Laterally flattened.** The blue-and-yellow striped angelfish seen here are more laterally flattened than most other fish. It allows them to retain their swimming speed and gives them the additional advantage of fitting into narrow crevices, whether seeking protection or food.

Their near relatives, the butterflyfish, are flattened this way too. The spadefish, a white-and-black striped schooling fish of the tropical seas, takes the same form.

**Fusiform and laterally flattened.** Soldierfish such as these are one of the many species that combine fusiform and laterally flattened shapes. They dwell in the open areas of reefs and come out frequently at night.

The squirrelfish, which are in the same family, share these characteristics. Another family of fishes that outwardly resemble them, the big-eyes, do too. Some of the small species of sea bass that never grow to more than four or five inches in length tend to be of this shape as do the tiny cardinalfish and several species of wrasses.

**Fusiform.** This blue-spotted rock cod, a sea bass found on Australia's Great Barrier Reef, comes close to the torpedo shape of fusiform fish. When it must, the sea bass can move at express train speed to strike its prey.

Other members of the sea bass family share the shape of this fish. So too do the billfishes, including the several species of marlin, the swordfish and the sailfish, which exceed the sea bass in speed. Lizardfish strike their prey with reptilian swiftness.

**Attenuated and cylindrical.** Trumpetfish, like this one, are attenuated in shape and cylindrical in general form. They are not as flexible as the eels or morays, but their proportions are rather eellike and enable them to mimic some of the staghorn corals they normally dwell among.

The rarer cornetfish resembles the trumpetfish strikingly, differing chiefly in having a long filament extending from the tailfin. Pipefish, who dwell in the eel grass and manatee grass of tropical and temperate coastal areas, match these plants with their long, drawn-out, and cylindrical shapes.

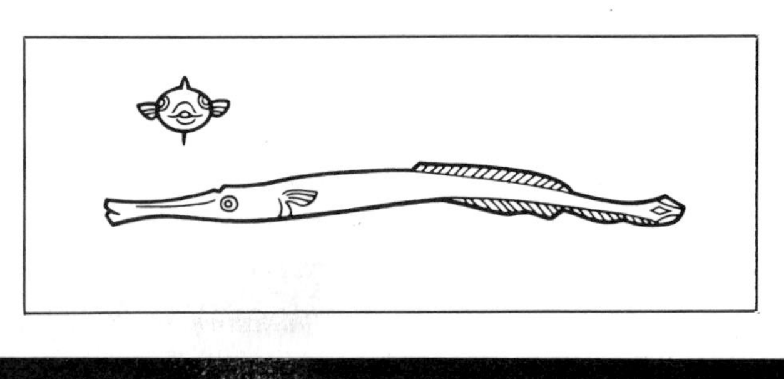

**Attenuated.** The moray pictured here, a variety of eel, has an attenuated shape, which enables it to slip in and out of openings in the rocky reefs it inhabits. Having a small diameter relative to its length, the moray is able to seek food in the small holes that honeycomb many reefs. It frequently lies in a dark hole waiting for a potential prey to pass by. Unwary divers have thrust hands into holes and have been bitten, but morays are usually not aggressive.

The morays are only one of more than two dozen families of eellike fishes in the same classification that have an attenuated body.

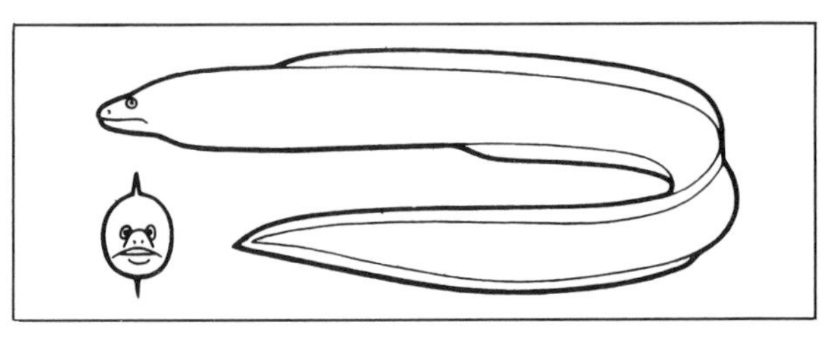

**Dorsoventrally depressed.** The flattened form of this stingray enables it to move along the bottom almost surreptiously. Flattened top and bottom, it is the epitome of the dorsoventrally compressed form. In adapting to life on the ocean floor, rays, and their relatives the skates, have become flattened and have developed a unique means of locomotion dependent on their body shape.

So too have their relatives the sawfish and the guitarfish. A shallow water angler, the goosefish, is another dorsoventrally compressed fish that dwells almost exclusively on the sea floor.

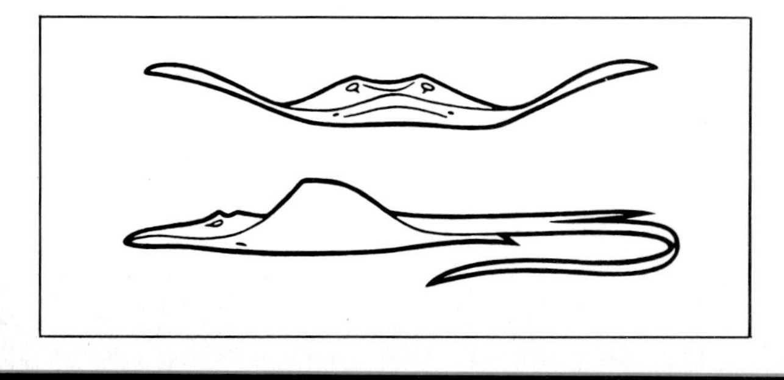

**Combining shapes.** These hardtails are members of the jack family, those swift-swimming relatives of the pompano that range the open oceans of the world. Their somewhat flattened fusiform shape and reinforced narrow tail base place them among the most powerful swimmers in the sea.

Other swift swimmers that fit this category include the tuna, mackerel, and mako. In addition the striped bass of the north temperate coastal waters and the bluefish of the same areas have a combination shape. Parrotfish of the tropical coral reef areas of the world do too.

## Cetacean Shapes

The cetaceans, which include the whales, dolphins, and porpoises, have become adapted to a totally aquatic life since their ancestors returned to the sea nearly 70 million years ago. There is little evidence of cetaceous ancestors, but most people consider them to have been omnivorous animals possibly like some hoofed animals today. Recent studies on the chromosomes of baleen and toothed whales indicate that there may have been more than one single ancestral stock. More closely related species have chromosomes which appear very similar, but among cetaceans there are some surprising variations.

The most important changes were those having to do with the way the animals moved and breathed. They reassumed the fusiform shape of early fish. The bones in their necks became shorter until there was no longer any narrowing between head and body. With water to support their weight, they became rounded or cylindrical in body shape, reducing the drag irregularities. Front limbs adapted by becoming broad, flat, paddlelike organs. But internally, cetaceans still retain evidence of fingers. X rays reveal up to 12 extra finger bones in some species. The hindlimbs disappeared, leaving only a trace internally that there ever were any. The tails developed into flukes. It is the flukes, combined with the powerful muscles of the trunk of these animals, that provide the propulsive power enabling them to swim and dive so efficiently. Unlike the fish's, a cetacean's flukes are horizontal, moving up and down. In fish tail fins or caudals are vertical and move from side to side, providing an easy-to-recognize differentiation between fishes and mammals in the sea.

Today's whales, dolphins, and porpoises are born, live their lives, and die in the sea. They

are totally aquatic. Their movements are appropriate only for the aquatic environment. Their streamlined forms and powerful flukes enable them to swim with ease. The big whales have grown to be the largest animal that ever lived, outweighing the biggest dinosaurs 3 to 1. Such large structures could not survive on land—because of their body weight they would suffocate.

Another change the cetaceans underwent in adapting to their reentry to the sea was the

> "These returned-to-the-sea mammals are voluntary breathers, breathing only upon conscious effort—unlike man."

position of their nostrils. From a position on the upper jaw as far forward as possible, the nostrils moved upward and backward until they are today located atop the head, sometimes as a single opening, sometimes as a double opening. And these returned-to-sea mammals became voluntary breathers, breathing only upon conscious effort—unlike man and other mammals who are involuntary breathers.

The development or return of a dorsal fin for lateral stability was another change that took place in some of the cetaceans upon their return to the sea.

***A / Porpoises.*** *Tapering profile and jaws integral with skull.*

***B / Sperm whale.*** *Enormous square head and heavy body.*

***C / Dolphin.*** *Tapering head and beaklike jaws.*

***D / Gray whale.*** *Heavy bodied with mouth high on head for filter feeding.*

***E / Narwhal.*** *Has a single upper incisor that resembles the mythical unicorn's horn.*

***F / Humpback whale.*** *Tapered in front and back, but thick-bodied between very long flippers.*

## Imitating Sea Animal Forms

***The racing motorboat,*** *above, a flat plane that skitters over the water's surface, seems to fly. Its pointed bow helps it knife through the water as it gathers speed to lift up out of the water. The shallow keel, which runs the length of the otherwise flat-bottomed hull, gives the boat stability.*

*At right, the* ***Atlantic dolphin*** *with its torpedo-like body speeds along, sometimes leaping clear of the water as it goes. The speed it attains enables it to make these leaps—like the motorboat that breaks free of the water. And it is during these hurdles that the dolphin can breathe without breaking stride.*

Man has taken many lessons from animals that inhabit the sea. Our imitations of animal forms show in aircraft, submarines, blimps, surface craft, and automobiles. We seek to imitate the streamlined form of fish like the tuna, the mackerel, the shark, and other fusiform fish which are built for speed. Airplanes and submarines, for example, have cylindrical bodies, tapering at the ends. Both have broad plane surfaces analogous with a fish's pectoral fins. In the airplane it is the wings. In the submarine it is the diving planes. Both have a vertical plane analogous to either the dorsal fin or the caudal fin of the fish. In the plane it is the vertical tail assembly. In the submarine it is the sail, that part of the boat that houses the periscopes and conning tower jutting above the pressure hull. Our surface craft are handicapped by the fact that they move half through water and half through air and that their design is a compromise between good streamlining and seaworthiness. But of course there is no model of a wheel in the animal world, and it is the privilege of man to have materialized the axis of a spinning body. Once the wheel was invented, it was inevitable that paddlewheels and screws

would be used to push various types of hulls through water. Modern ships' propellers are really just an assemblage of several rotating fins. Slow ships like tugs have huge propellers, each blade having hydrodynamic characteristics comparable to those of a grouper's tail, while speedboats have small, rapidly rotating screws, with short and broad blades like the caudal fins of tuna.

Comparing all animals and all man-made vehicles moving through air, on land or underwater, the most efficient will be those that require the least energy to transport one pound of matter over one mile. By far the best result is obtained by man on a bicycle, second best are open-ocean fish and sea mammals, then horses, then jet airplanes. Way behind come helicopters; bees and mice trail the list. Ironically, the simple bicycle is the only invention of man that beats marine animals in efficiency.

# Chapter III. Traveling Wave

When a fish hanging motionless near the bottom is disturbed, it bursts from its resting place with only a few rapid strokes of its tail, fleeing possible trouble. It would seem that a fish should have great difficulty accelerating rapidly in a medium as dense as water. Some of the energy is spent to generate acceleration; the rest is spent to counter friction and pressure drag, which are roughly proportional to the square of speed. Yet fish seem able to reach their maximum speed by flicking their tails a few times. To accomplish this feat, they must be extraordinarily able to get good "traction" on the water around them. How can they achieve the great thrust necessary for quick movement?

**"Fish reach their maximum speed by flicking their tails a few times. How do they achieve this great thrust?"**

The most common form of locomotion among the aquatic animals is undulation. The body is thrown into a series of curves that begins at the head and passes along the length of the body as a *traveling wave.* Among most fish these body waves move in the horizontal plane, but in flatfish and in many marine mammals the body waves move in the vertical plane. One of the most typical examples of pure traveling-wave propulsion is an aquatic reptile, the sea snake, quite common in some equatorial seas.

The earliest vertebrates probably swam in a similar manner, with rhythmic muscle contractions flexing their bodies from side to side. Pushing against the water this way resulted in forward movement. Consistent with this theory is the primitive lamprey, or hagfish, which uses the traveling wave.

Nearly all fish employ one of three general swimming techniques stemming from these traveling-wave movements. The first is that employed by the sea snake, and attenuated fish, like eels and ribbonfish. Their motions are serpentine; the traveling wave bends their bodies into curves that increase in amplitude and decrease in wavelength as they progress backward. Because they have only small ineffective swimming fins, if any, they entirely rely on undulations of the body to gain a "foothold" on the water.

Fish with rigid bodies, on the other hand, like the armor-plated cowfish, are unable to flex their bodies to help them swim. They use only their short tails, swishing from side to side in a short arc, to push them along.

Between these two extremes is a third method, used by the majority of fish, combining characteristics of both. The wigwagging of the caudal fin is coordinated with subtle body undulations. This method yields smooth, rhythmic motion and improved efficiency. The fish's head swings in a small arc, and the body and caudal fin curve to form a complete transverse wave. So our startled fish, if he belongs to this category, gets a powerful takeoff leaning on the water with its body as well as with its tail. Men with their rubber foot fins do their best to imitate this type of swimming.

***Epitomizing the traveling wave.*** *Startled into action, the sea snake gives a perfect example of the traveling wave. Distinct S-curves travel from the animal's head and along its flattened body, and in increasing breadth, finally reach the tail. Studies of the eddies created at each curve of the body indicate that the supple snakes may actually roll on the swirls they cause in the water.*

***A spotted trunkfish** sculls about the coral reef in search of food. Its rigid body makes it a clumsy swimmer, but in an emergency and for a short time, it is capable of surprising bursts of speed.*

## The Living Box Moves

The inflexible armored body of the trunkfish is fine protection against some of the predators it must occasionally face. But it renders the strangely shaped fish awkward and slow moving. In cross section, the trunkfish and its near relative the cowfish are either triangular or rectangular, and their shapes further contribute to their awkwardness. These animals in armor are poorly equipped to outswim an enemy. Under normal circumstances, with no cause for alarm and no imminent danger, the trunkfish sculls about with its transparent, fan-shaped pectoral fins. Its equally small, delicate dorsal and anal fins contribute a wave form motion in this normal mode of swimming. Its caudal fin barely moves. Frequently, the fish hangs in the water on the reef, barely moving any of its fins. It remains near the bottom or close to cover of some sort. When danger approaches, the awkward trunkfish and cowfish flail about to reach cover. Their caudal fins lash violently from side to side on the hinge at the base of their tails—the peduncular hinge marked with a sharp spine that offers additional protection against attack from behind. But the main thrust of swimming comes from a special way of moving developed over many thousands of generations. The accompanying photograph and chart show how the trunkfish and cowfish move to meet these occasional life-and-death emergencies.

This unique method of swimming is called ostraciiform movement. It is peculiar to the trunkfish, the cowfish, and two other fish families that are related to them, the puffers and the porcupinefish. The name for this type of swimming comes from the family name (they are variously called Ostraciidae and Ostraciontidae) of the trunkfish and the cowfish.

In the trunkfish and the cowfish it is the bony armor that makes ostraciiform swimming necessary. Among the puffers and porcupinefish, which use the same movement, the need arises from their awkward shape.

The trunkfish's bony armor is best described as a living box of polygonal bony plates. Some of the plates have minute spines and tubercles giving their surface a rough feeling. One curious habit some divers have observed in trunkfish and cowfish is that of blowing jets of water at the fine sediments on the sea floor around the reefs. They probably use these jets in their search for the minute organisms they eat. Each time they use these water jets, the sand in front of them billows up in clouds bringing up a few tiny organisms. At the same time, the jet pushes the fish back an inch or two. The visual effect is a back-and-forth movement which is interrupted periodically when the trunkfish (or cowfish) darts forward a few inches more than usual to suck up the small animals it has found.

***A / Ready for movement.*** *By contracting all musculature to one side, the trunkfish puts its somewhat streamlined head and its tail fin on the same side of its median axis.*

***B / Movement begins.*** *The fish now straightens itself and its tail and head cross the axis of progression at the same time. At this instant the tail has a strong forward thrust, and this drives the animal ahead. Pausing in this position momentarily, the fish takes advantage of the flow of water that pushes it along.*

***C / Movement continues.*** *Once again the trunkfish contracts all its muscles, this time to the side opposite that in position* ***A****, with head and tail together on the other side of the median axis.*

***D, E, and F / Repeating the process.*** *The fish continues the movement, alternating from one side to the other, and makes its way through the water.*

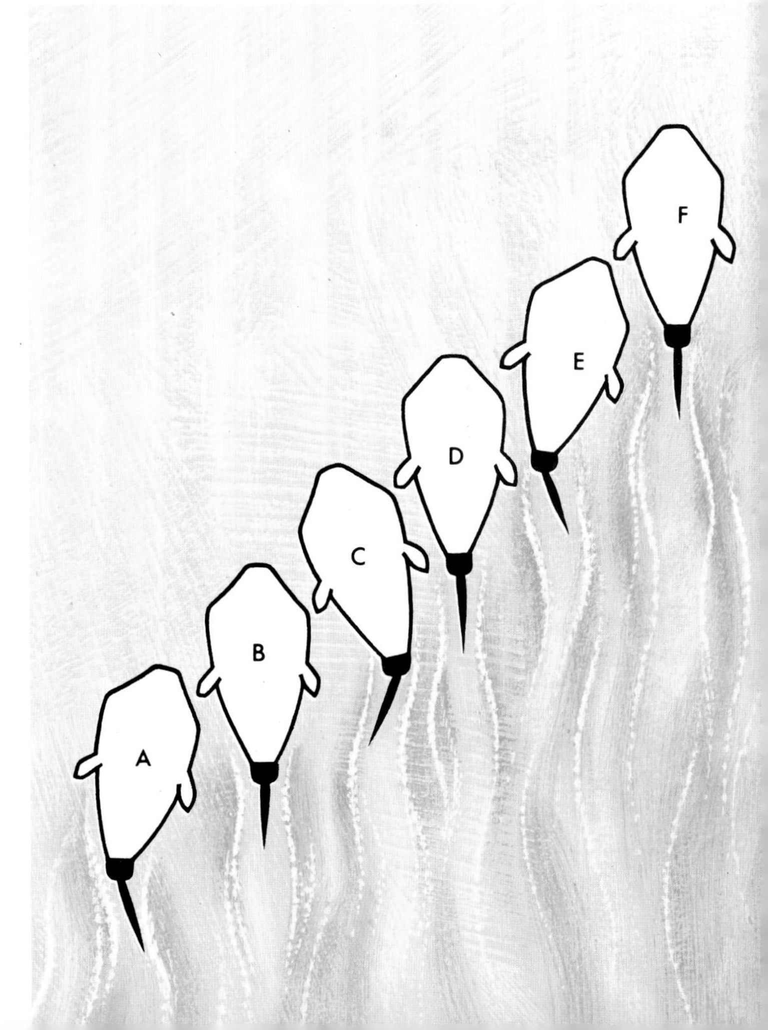

## Effect of Muscle and Bone Action on Swimming

It was once thought that fish swim solely by using their tail fins, which they swept in an arc behind them. But motion-picture analysis of fish shows that fish swim by side-to-side undulations of their entire bodies. In fact, fish whose caudal fins have been amputated can still swim, some almost as well as intact fish of the same species. We know the tail fin facilitates swimming, but it is not the exclusive source of propulsion power for most fish.

The muscles along the sides of a fish are the strongest it has, and those associated with the fins are relatively weak. In swimming, a succession of contractions passes along each side of the fish. Some of the W-shaped muscle segments, or myomeres, contract on one side, and those opposite them relax and

stretch. This bends the fish's body, and the fish pushes against the water first with one side and then with the other.

The flexible frame of the fish is a good foundation for the muscles. The backbone extends from the head to the tail and is made up of many interlocking vertebrae. The vertebrae are jointed to allow side-to-side movement and are strong enough to withstand the great strain placed on them by the flexing muscles.

***A / Cutaway of a salmon.*** *This shows the backbone, or vertebral column, of the fish and some of the numerous bands of muscle segments that gird the fish. Note that the spiny rays of the dorsal and anal fins are not attached to the backbone but simply are anchored in the flesh of the fish.*

***B / The backbone bends from side to side.*** *Muscle contractions pass in waves along each side of the fish. While a series of muscle segments contract on one side, those on the opposite side relax, allowing the fish to bend at that point. The fish gets its thrust when it contracts and relaxes muscles alternately—first on one side, then the other. So the principal thrust comes not only from the caudal fin itself but also from these muscle segments or myomeres.*

## Orca's Traveling Wave

In an easy undulating motion, the orca (killer whale) breaks the water's surface with the top of his head. A puff of vapor issues from his blowhole as he exhales. As he quickly inhales, his back with its tall dorsal fin breaks the surface too. He bows his head and points downward even as his great, broad flukes flash momentarily out of the water. And he's gone—submerged beneath the sea's surface as smoothly as if he hadn't passed that way with his multi-ton hulk.

Those great broad flashing flukes on orcas and other whales and dolphins are planes that normally are parallel to the surface of the water. They are moved up and down driven by powerful muscles in the body ahead of the tail. The muscles are connected to the tail and flukes by a series of tendons. When the orca bows its head downward to sound, its back arches and a traveling wave passes along the length of the animal's body with increasing amplitude. As it reaches that part of the body that houses the muscles that power the tail and flukes, the great ani-

***Surfacing.*** *Bubbles of exhaled breath appear as the orca surfaces to take some fresh air.*

***Fin over surface.*** *The dorsal fin appears above the surface with water streaming from it.*

mal "snaps the whip" and the broad flukes drive the animal forward.

It is a fortunate coincidence that the vertebral column of marine mammals developed on land by their ancestors allows them to bend in a vertical plane, so that dolphins and whales can "sound" to feed and "surface" to breathe easily and often. Fish, on the contrary, which do not need to surface, have developed laterally flexible spines.

In captivity, orcas have demonstrated some of their vast strength. They can learn to leap high up, leaving the water entirely with their great bulk. In some aquariums where they are held, the orca must share a pool with a companion, a white-sided dolphin. Although the dolphin is dwarfed in size it is apparently compatible with the ineptly named killer whale. These dolphins move in virtually the same manner the orcas do. Some can "tail walk," leaping up until only their flukes are in the water, then remain in a vertical position just above the water by wagging their flukes vigorously just beneath the surface.

***Sounding.*** *Arching its back, the orca prepares to sound. Its head has already entered the water.*

***U-turn.*** *Dorsal high in the air, the orca is in the middle of a sweeping vertical U-turn.*

# Chapter IV. The Role of Fins

Fins are to fish what arms and legs are to men. And even a little more. Most fish have two general types of fins—median, or vertical, fins, which originate along the midline of the animal, and paired fins on their sides. The median fins include the dorsals on the back; the caudal, or tail, fins; and the anal fins on the belly just behind the vent or anus. Some fish have as many as three dorsal fins; some have two anal fins. Of the paired fins, the pectorals at each side near the head are analogous with our arms and with bird's wings, and most fish have them. The paired pelvic, or ventral, fins are located below and usually behind the pectorals.

Fins have been adapted for many purposes, but they are mainly used for propulsion, stability, steering, and braking. In some cases, as we'll see later in this volume, they have been modified for other functions. As fish evolved toward faster speeds, their fins, like feathers on an arrow, provided stability and made it possible for them to propel themselves where they wanted to go. As its head moves from side to side when a fish swims, the animal has a tendency to veer off from its forward path. To resist this condition, known as yaw, the fish erects its dorsal fin. The tendency is further reduced in fish with long, slender bodies and by the long, trailing dorsal and anal fins of some reef fish with deep, short bodies. These deep bodies keep the fish from rolling, much in the way a sailboat's keel keeps it steady. Fish of other body structures extend their paired fins to avoid rolling. To change direction vertically, fish bring their paired fins into play. The pectoral fins act as hydroplanes to raise the nose of the fish, while ventral fins bring the rest of the body into horizontal plane. All these fins may be modified for other functions depending upon the various species. For some the dorsal, anal, or pectoral fins are the agents of propulsion, while in others the body shape necessitates long flowing fins for stability.

> **"A long and soft tail gives the grouper instant thrust. The broad and short caudal fin of tuna is a high-efficiency propeller for fast, long-range travels."**

The main propulsive agent for most fish is the caudal fin. A broad tail gives a rapid burst of speed from a standing start, and this is useful to a fish that must dart after a meal or away from a predator. Fast, long-distance swimmers have a very long and narrow caudal fin, which is hydrodynamically very efficient and seen in such fish as tunas, jacks, and marlins. Laboratory experiments show that for a given fish the amplitude of tail beats remains practically constant and that speed depends primarily upon the number of beats per second, which have been measured as high as 25. Seahorses and some species of eels get along without caudal fins, and in skates and rays they have developed into long, spikelike projections with virtually no function in propulsion.

***Maneuvering for a turn.*** *This rockfish, a member of the scorpionfish family, looks like it's signaling a turn as it extends a pectoral fin. Actually the fish is using its pectoral fin to maneuver into a turn. Giving a major assist in the turn is its tail fin, curved off to its left. The soft-rayed rear part of its dorsal, or back, fin is also curved to the left, perhaps helping in the maneuver, while the spiny-rayed portion of the dorsal remains erect, lending stability so the fish won't flop over on one side. Turning is only one of the many functions of the various paired and median fins fish possess.*

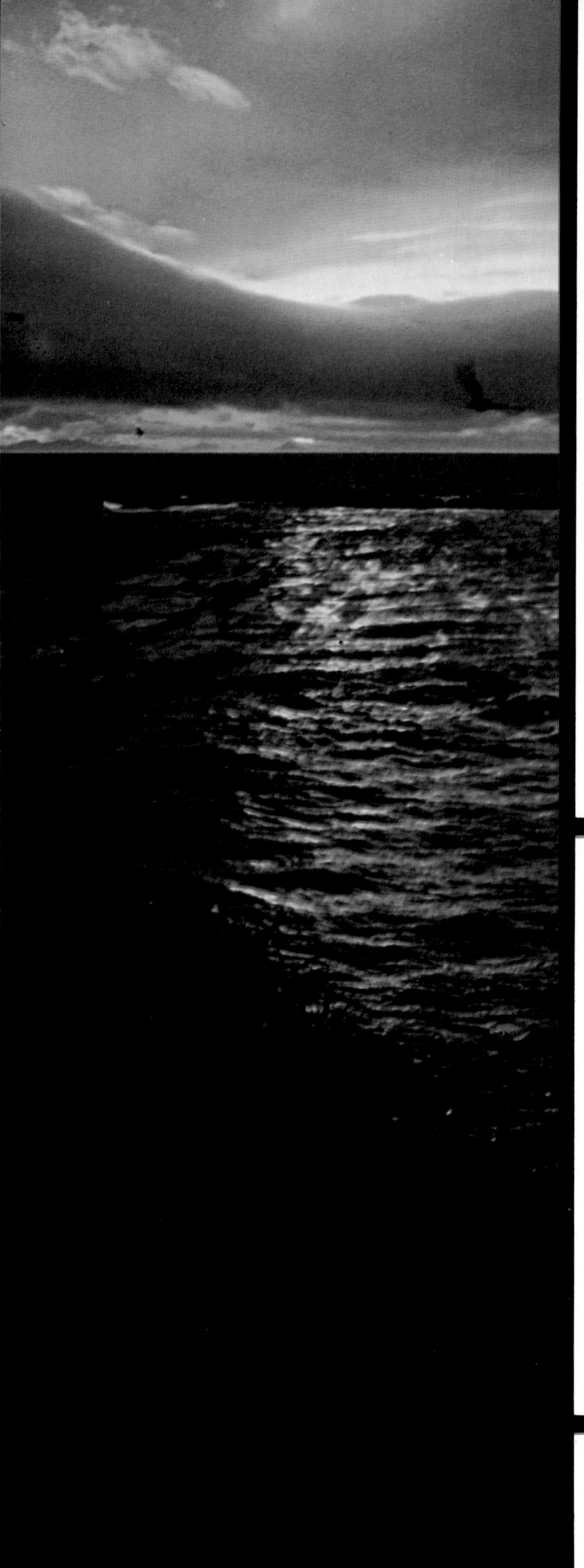

## The Important Propulsive Fin

Whatever the shape of the caudal fins, their function is essential for aquatic animals. The caudal fin, or tail, is used in coordination with the massive muscle segments of the body to propel the fish through water. Most fish tails originate at the end of the vertebral column. In heterocercal (uneven) tails, the spinal column extends into the larger upper lobe of the caudal. The heterocercal tail of sharks gives these heavy fish without swim bladders an upward thrust. In homocercal (even) fins the vertebral column ends at the fin base and supports a symmetrical "tail" which produces only a forward thrust. Some homocercal tails are forked, some are square, some are rounded—but each serves a specific function. For example, the lunate (crescent-shaped) tails of mackerels, tunas, and jacks indicate fast swimmers. The broad tail of a grouper gives him the ability to accelerate very quickly.

***Tenpounder.*** *This herringlike fish can change its width for different propulsive demands.*

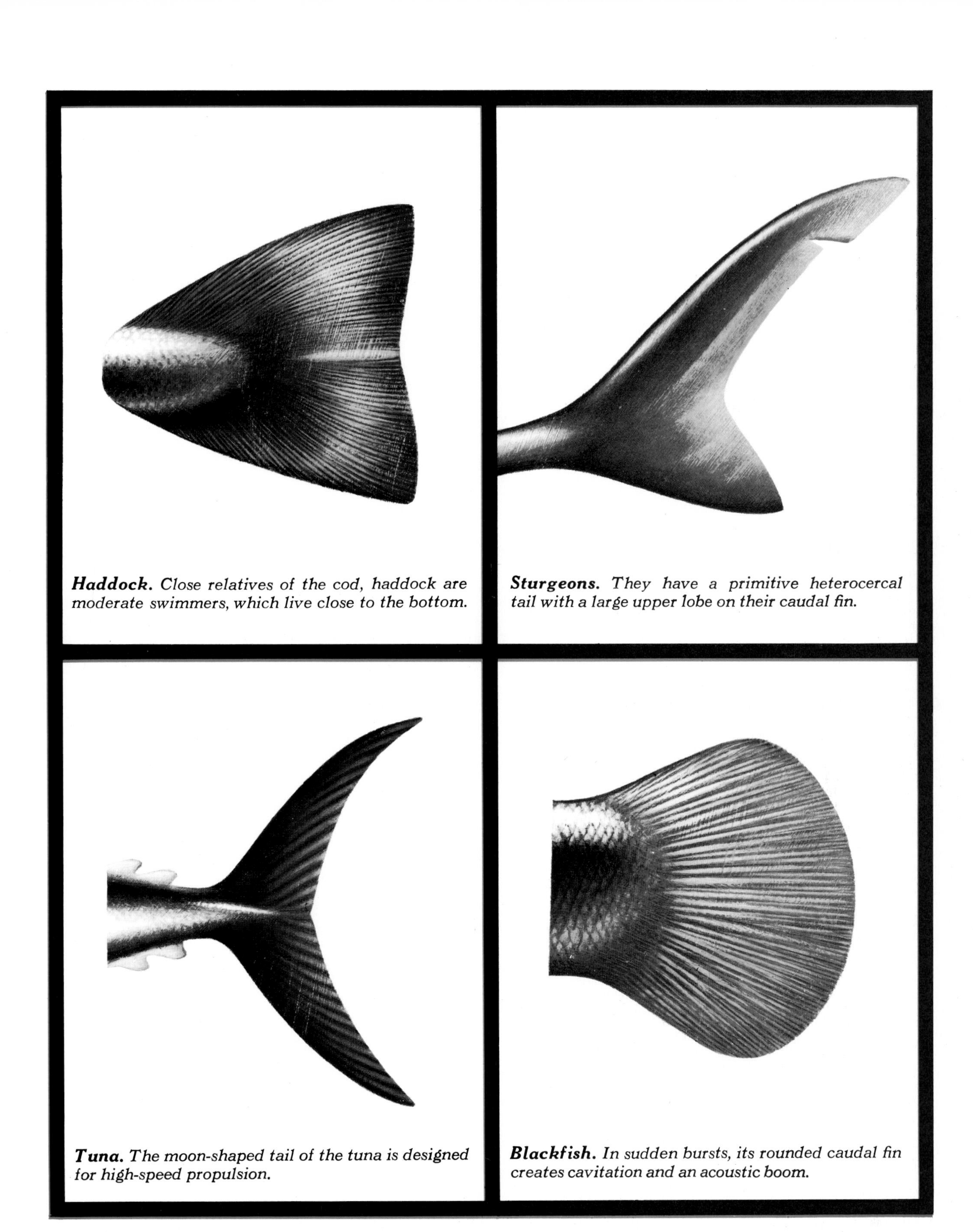

***Haddock.*** *Close relatives of the cod, haddock are moderate swimmers, which live close to the bottom.*

***Sturgeons.*** *They have a primitive heterocercal tail with a large upper lobe on their caudal fin.*

***Tuna.*** *The moon-shaped tail of the tuna is designed for high-speed propulsion.*

***Blackfish.*** *In sudden bursts, its rounded caudal fin creates cavitation and an acoustic boom.*

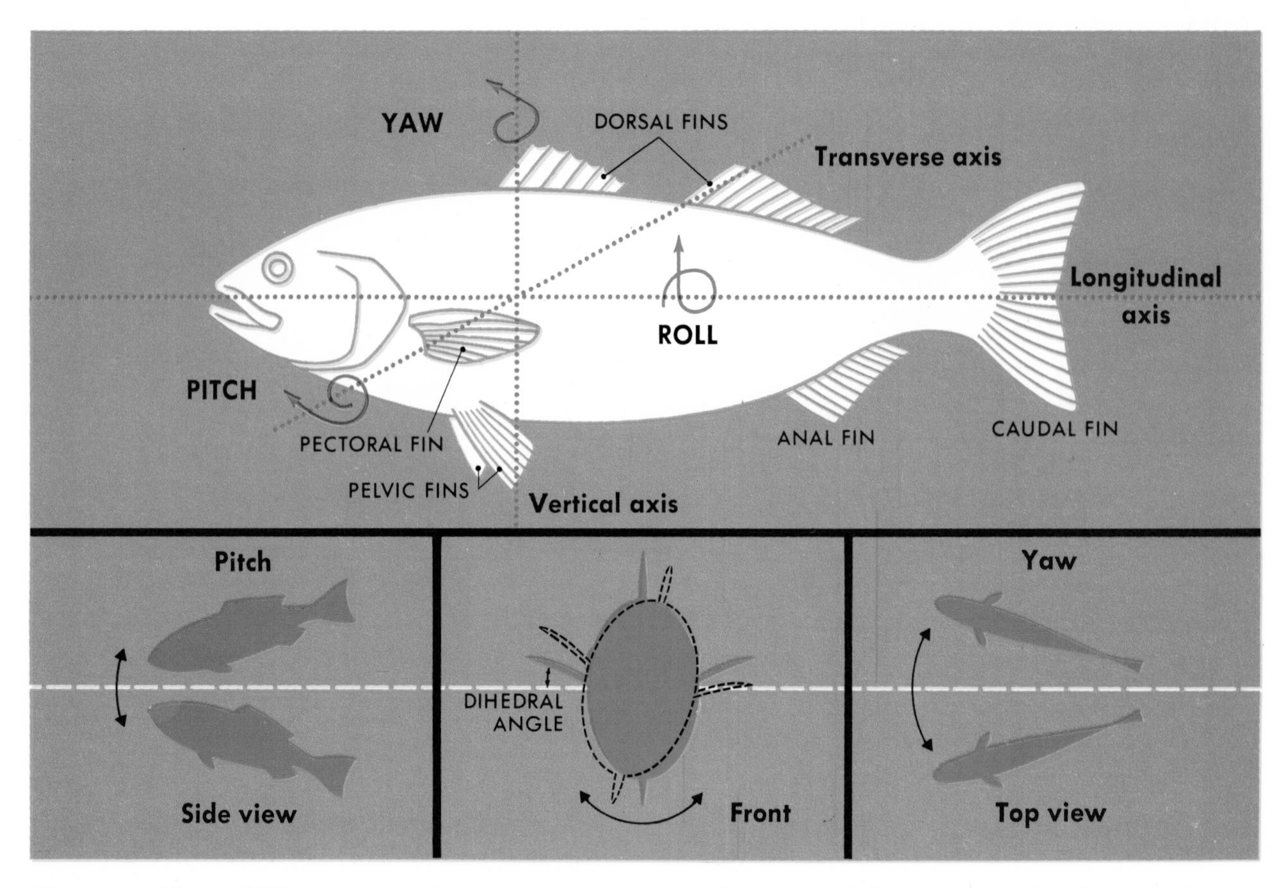


***Fins provide stability.*** *Fins of fish resist movement about the animal's body axes, overcoming the tendency to pitch, yaw, or roll. Upswept forefins (dihedral) prevent roll.*

## Counteracting the Sea's Forces

The world of the fish is three-dimensional with forces pushing and pulling in all directions. These forces are gravity, which tends to pull the fish downward; buoyancy, which tends to hold it up; and drag, which tends to hinder its forward motion. Secondary effects of these forces are pitch, roll, and yaw. In pitching, the fish's head rocks up or down on its transverse axis. To counter pitch, the pectoral fins extend at the proper angle of attack and camber. In rolling, the fish tilts from one side to the other about its longitudinal axis. To resist roll, all the fins are extended. Yaw is the tendency of a fish to turn about its vertical axis. To resist yaw, the unpaired dorsal becomes erect as well as the anal fins. Maneuverability requires the ability to brake rapidly, then all fins and the tail are curved to offer maximum resistance.

***Moorish idols.*** *The laterally compressed body of these disc-shaped Moorish idols gives them great lateral stability to prevent rolling. And their tail, dorsal, and anal fins give them even greater ability to control side-to-side yaw. This ability stands them in good stead when they dart among Pacific coral reefs, where the surge of waves could dash them against the jagged coral. The fish's narrowness also allows it to delve into tight little corners of the reefs for food. To maneuver in and out of tight coral crevices, the Moorish idol is capable of turning in less than one-half the length of its body.*

## Fins to Lift, Stabilize, Stop

These requiem sharks, built for speed, have difficulty stopping. Unable to brake themselves effectively, they must make sharp turns. Because of this, few sharks venture in coral reefs where space is restricted, and they roam around the reefs. Their fins are used for maneuvering.

***This white-tipped reef shark*** *(above) glides along on its pectorals, using them as if they were wings. The shark's large, white-tipped first dorsal and tiny, dark second dorsal, along with its anal fin, all combine to give it lateral stability.*

**A Carcharhinus *shark*** *(right) uses its pectorals, twisting one up and the other down, to effect a sharp turn, perhaps to slow down. The combined action of its caudal fin with its pectorals helps make the drastic maneuver successful. The shark's dorsals keep its body from rolling.*

***A sea bass** hovers quietly by gently sculling while a small cleaning goby picks parasites from its mouth.*

## To Start, Hover, Turn, Stop

The basic simplicity of design of a typical fish or of a marine mammal, when compared to that of a lobster, of a giraffe, or of a man, is obviously due to neutral buoyancy and to the absence of strong, large, complex limbs. Thus the muscular structure can be concentrated in one solid pack. We have seen that fins other than the tail are used for fine maneuvering, which enables the fish to master its liquid environment. By using the proper fins in the proper ways, they can stop in midwater or they can hover in one spot, or they can turn quickly and start moving with a sudden burst of speed or at a leisurely pace.

***Nassau grouper*** *brakes by fanning its pectorals out to full width and turning them to face forward.*

In making these various adjustments they depend largely on their pectoral fins, which they coordinate with the movement of their other fins. Sea basses, like most other bony fishes, have swim bladders by which they can regulate their buoyancy. Working together with the fish's fins, this buoyancy control enables them to move about in almost any conceivable manner through their three-dimensional world—up or down, forward or backward, left or right.

Many divers in tropical waters have seen groupers, those large members of the sea bass family, hovering patiently near them, sizing them up.

## Compressed Bodies

***Yellowtail tangs,*** *seen here in a school, offer little resistance to the water they swim through because of their slender shapes, streamlined in horizontal sections.*

The compressed bodies of yellowtail tangs give them stability by preventing any tendency to roll while enhancing their ability to make sharp turns. They generally move in unison in schools. Even their caudal fins are turned the same way as they move compactly through the water. Tangs, and their relatives the surgeonfish, swim with beats of their weak tails and by "sculling" with their pectoral fins. Their slender bodies knife through the water and offer potential predators a shape difficult to swallow unless it is seized just the right way. Other advantages of the slender shape to the tangs, surgeonfish, and other species that have compressed bodies include the ability to slip into narrow crevices in rocks or coral reefs to seek shelter against their enemies.

## Living Metronomes

Pufferfish, triggerfish, and ocean sunfish move about in a unique way. Their bodies remain practically rigid and they move by waggling dorsal and anal fins synchronously to the same side, which gives them a readily recognizable appearance. When undisturbed, they leisurely beat these fins left and right like living metronomes.

Though dorsal and anal fins are in charge of most of the triggerfish's routine propulsion, powerful strokes of their tails enable triggerfish to make sudden "spurts."

***This clown triggerfish,*** *as well as other members of the triggerfish family, are not especially swift swimmers, and may have difficulty outrunning predators. Perhaps to offset their relative slowness, the first spines of their dorsal fins have been modified into a movable unit that acts as a unique defense. The first spine, being longer and thicker than the rest, can be erected by muscles then locked into place by the second. Until this spine is lowered the first cannot be depressed. When raised, the triggerfish's spine makes it difficult for enemies to eat.*

***Moray eel.*** *Here a brown moray lies on its right side while its head stands straight. Morays occasionally lie upside down.*

## No Lateral Stability

Morays are eellike fish that have no pectoral or pelvic fins. They do have long, low, fleshy dorsal and anal fins, which meet and unite with their caudals. They move by undulating their bodies and those long median fins in a traveling wave. Living as they do in rocky reef areas, they don't need paired fins. Their lack of a high dorsal denies them much lateral stability, and it is not at all unusual to see morays lying on their sides even if nothing is wrong with them. They don't need the lateral stability that large dorsal and anal fins offer, because most of their lives is spent resting on the firm base of rock walls, rather than floating in midwater.

They do not actively seek out trouble, but a diver putting his hand into a hole with one of these creatures could receive a nasty bite.

***Inflation.*** *This pufferfish inflates with water to protect itself against an approaching predator. That predator may find the inflated puffer too big to swallow or too imposing in its new size.*

## Dependence on Fins

The pufferfish moves principally by sculling with its pectoral fins. Its small dorsal fin and its caudal fin come into play to some extent, wiggling rather feebly from side to side. When a puffer inflates itself with water, its body becomes even more stiff than it is normally. In such a hopeless situation, it cannot depend on even a slight body motion to help it move, but must rely entirely on its fins.

Because the puffer is so slow moving, it depends on its ability to inflate itself for protection and in some species on sharp-pointed spines that become erect when the fish swells. Its coloration, which tends to blend with the reef environment, also helps it to survive predation. Unfortunately, man can cope with these defenses and puffers and porcupinefish especially are often caught and dried to be sold as curios.

Two additional defenses puffers have are hardly visible. One is their ability to bury themselves in sandy ocean bottoms by squirming. The other protection is the toxicity of the fish for those eating them. The poison is concentrated in the liver, the viscera, and the gonads, especially around spawning time. Because some species of puffers (the fugu) are considered a delicacy on the Japanese table despite their poisonous qualities, cooks preparing them for human consumption often must show proof of graduation from a special school that trains them in detoxifying puffers. Seven thousand tons of fugu are eaten in Japan each year, but the emperor is not allowed to enjoy that dangerous delicacy.

## Large Fish with Little Tail

The ocean sunfish, or *Mola mola,* is big on dorsal and anal fins and very small on caudal and pectorals. In fact, the tail fin of the sunfish is so small that at times it looks as if the fish has none at all. The sunfish looks as if it has been cut off just behind its head, which accounts for another of its common names, the headfish. To shift its ponderous body, it uses its large dorsal and anal fins, moving them from one side to the other at the same time or to alternate sides. Its tiny pectorals fan out and help a little in steering. Sunfish are often seen at sea on summer days, lolling around on their side near or at the surface. No one knows where they spend most of their lives.

They are not eaten by humans, but sea lions bite off all their propulsive fins and store them, alive but helpless, on the bottom as food reserve.

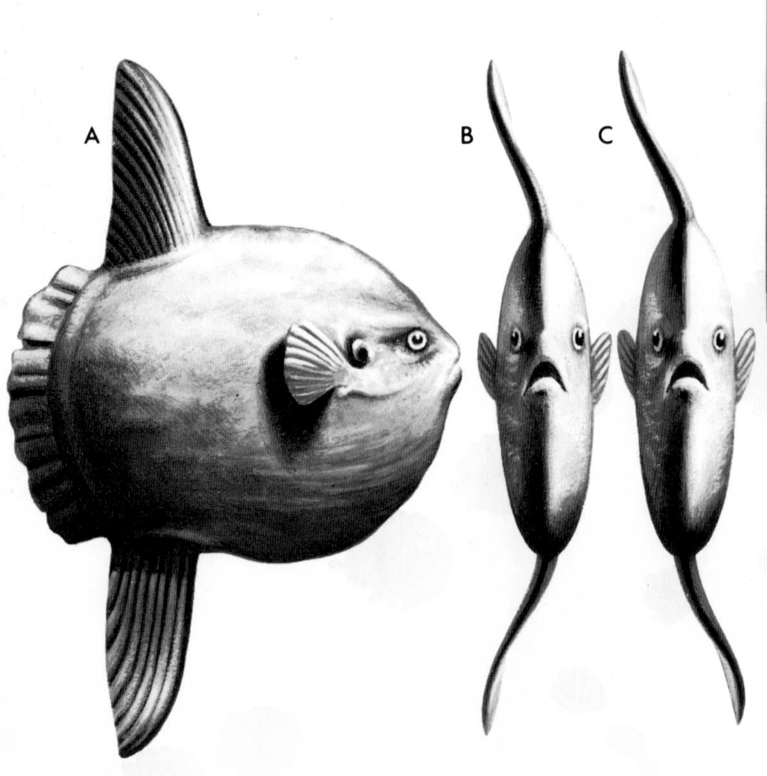


***A / A useless tail.*** *Since the tail of the ocean sunfish is so small, it is virtually useless. Instead the fish relies on its large dorsal and anal fins for propulsion. The dorsal fin occasionally sticks out of the water. But when this happens, it ceases to be of much use to the fish.*

***B / Waving its fins.*** *The dorsal and ventral fins are usually waved back and forth, each traveling to the same side of the fish at the same time.*

***C / Opposite fin movement.*** *If the dorsal fin is moving in air, the dorsal and ventral fins may get out of synchronization.*

## Undulating Fins

The seahorse's fins are nearly invisible, but close observation shows that the animal has control of each individual ray. High-speed photography has revealed that each ray is capable of moving at a rate of 70 times a second in an action similar to slats falling in a sagging picket fence. As an undulation passes from one end of the dorsal to another, the seahorse moves forward or backward, or up or down, in its own peculiar, very slow but versatile version of a traveling wave.

Being a poor swimmer, the seahorse usually avoids any areas where strong currents occur. It has sharp eyes and sits with its tail wrapped around seaweeds or gorgonians. When it does choose to swim, it unwraps its tail, straightens itself out, and begins to flutter its dorsal fin. Even so, these slowpokes may take one and a half minutes to cross a one-foot area.

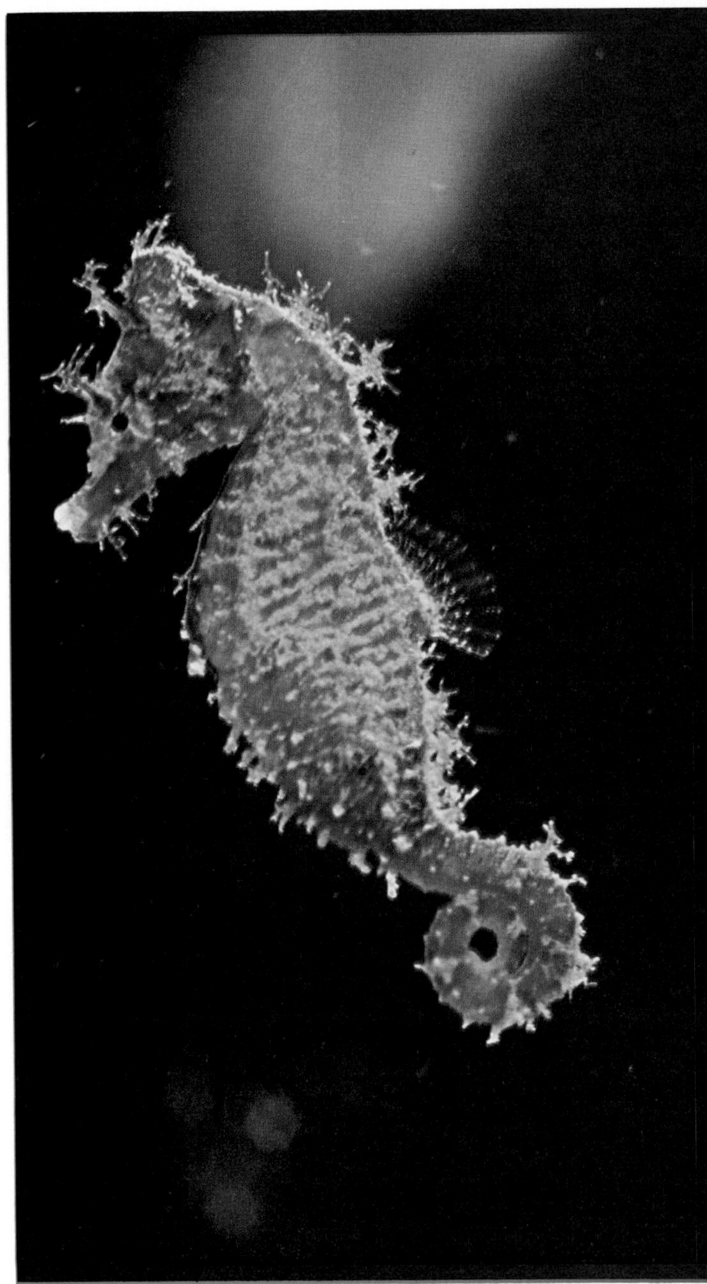

***Seahorse—sparse fins.*** *Looking very much like a chess knight, the seahorse "stands" in the grassy flats that are its usual home. Frequently it grasps a blade of eel grass, manatee grass, or turtle grass with its prehensile tail and remains stationary. But when it does move, perhaps to sip in some tiny zooplankton drifting by, it does so by vibrating its delicate dorsal fin as many as 70 times a second with a wave action that ripples through the fin each time it vibrates. Seahorses have no pelvic fins, a very small anal fin and no true caudal. Only a pair of pectorals, which help somewhat, and the transparent dorsal, which does most of the work, move the strangely shaped, vertically oriented seahorse about.*

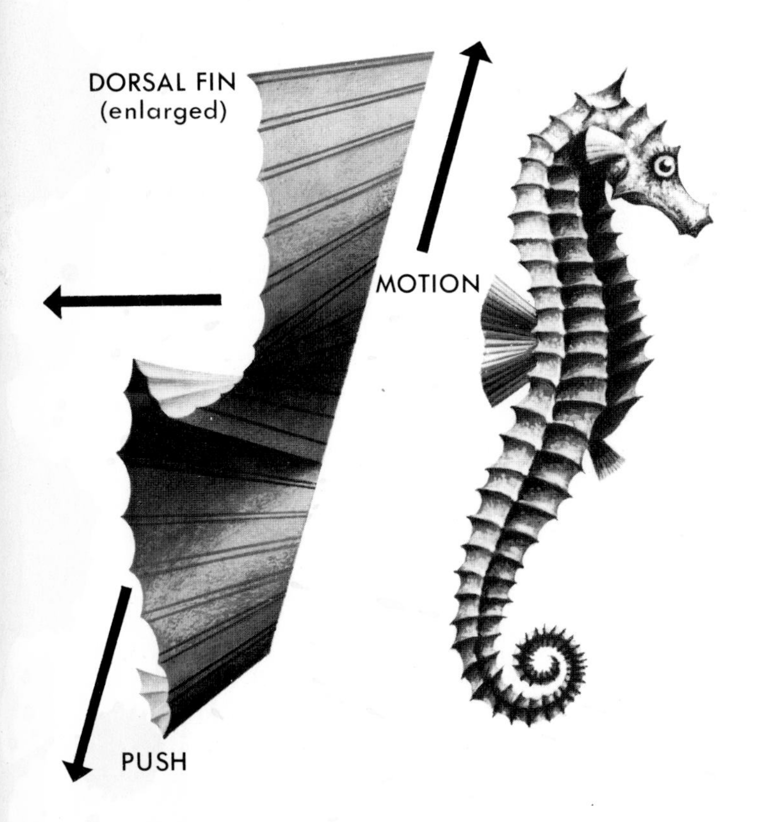

# Chapter V. The Jet Set

When a child blows air into a rubber balloon and lets it go when it is well inflated, it will fly about the room erratically as long as the pressure of the air inside the balloon is higher than that of the surrounding environment. This is because a certain mass of air streams out through the narrow neck of the balloon with a force proportional to the difference in pressure, and according to Newton's third law, for every action, there is an equal opposite reaction. It is the reaction that propels the balloon as air is ejected through the rear neck.

**"Fish that normally swim by more ordinary methods sometimes use a jet-assisted takeoff."**

The same experiment can be made in a bathtub with a hand-held shower nozzle. When the water is turned on, the shower head is pushed in the direction opposite to the flow of water. If the shower head is raised above the surface, the water jet is expelled into the lighter air, and there is a noticeable increase in thrust. Such thrust would even be slightly bigger if the water jet was expelled in the vacuum of outer space, which indicates that the efficiency of jet propulsion is handicapped in a dense medium. In the sea, jet-propelled creatures are capable of swift rushes or very slow cruising, but could not perform rapid, long-range migrations like tail-propelled animals.

In the world of the sea there are animals that move about in much the same way as the balloon, making use of jet propulsion. The simplest are salpa and jellyfish. Some of the bivalves, or two-shelled molluscs, accomplish jetting by very sudden contractions of their muscles. When the shells clamp shut, water is forced out and the animal travels in the opposite direction of the stream.

Jet propulsion is used by cephalopod molluscs like the octopus and squid. These animals energetically contract their mantle, forcing a narrow stream of high-pressure water out through a flexible siphon. The flow of the stream can be directed by the siphon to steer the creature with precision.

Fish that normally swim by ordinary methods sometimes use what is called in aeronautics "a jet-assisted takeoff." This is accomplished by the forceful opening and closing of their gill covers. As the covers snap shut, water is squeezed out and the fish gets an additional thrust forward.

Water-jet propulsion is now used by engineers to move boats in shallow water where propellers could get damaged, as well as for "bow thrusters," transversal jets located across the bow to increase dramatically the maneuverability of service ships or harbor tugs. Water-jet propulsion was also chosen for the first exploration submarine called "diving saucer" in spite of its low efficiency, because it offered an unmatched potential for maneuverability.

***This medusa,*** *a jellyfish, is a free-swimming form. It jets by alternately contracting and relaxing its bell-shaped umbrella. It is one of the cnidarians, some of which are jet-setters. Medusas are not very strong swimmers; they usually float along with the currents and the winds. But when they do move, it is with jerky, uneven motions. By contracting their umbrella, they force the water under it out the bottom. Normally this drives them upward. But they can also descend and move laterally. When they relax, the umbrella opens and admits a great volume of water, which is forced out with the next contraction. A primitive and inefficient way of moving, but it is sufficient for the jellyfish's simple needs.*

***Scallops.*** *These many-eyed, many-tentacled animals move by clamping their valves together, which causes a stream of water to shoot out and move them forward.*

## Escape by Dance

Scallops usually lie about on the sea floor in great numbers, quietly filtering from the water the tiny organisms that make up their diet. There they live, eating and growing and moving about only rarely. As protection against possible predation, they each prepare for themselves a small depression in the substrate on which they rest. Commercial fishermen seeking scallops locate these beds and drag special nets along the bottom to catch them. Even this sort of threat doesn't arouse the scallop to very much activity beyond feeding and pumping.

If one of the species of sea star that especially prey on scallops shows up in the vicinity of a scallop bed, it turns into bedlam. The scallops can sense their presence by chemical clues the sea stars inadvertently leave in the water. And these chemical clues are carried through the water to the scallops. The shellfish react by dancing away in a quick burst of jet propulsion, which more often than not leaves the predator some distance behind in a cloud of silt.

If one of the many species of sea stars that do not prey on scallops happens by, there is little or no reaction from the scallops. The chemical messages they send are not feared by the scallops. To escape from the predatory varieties, the scallops clap their shells together suddenly by sharply contracting their adductor muscles. Water is forced out from between the shells in twin jets.

## Primitive Jet Propulsion

Salps are simply structured animals with a complex way of life, which has an important influence on how they move. Related to the sea squirts, salps are among the most primitive of animals with a notochord, a sort of precursor of the spinal column in vertebrate animals. Many of the salps live not as colonial animals but in aggregates of many individuals having built great chains through the budding method of reproduction. Many of them are bioluminescent—giving off a greenish-blue glow. Each has an incurrent siphon to take water into itself; and each has an excurrent siphon to expel that water. Salps also have bands of muscles, which show plainly through their transparent exterior. They take water in and expel it by alternately contracting these bands of muscles. When they force the water out, they are forced in the opposite direction from that of the expelled water. Jet propulsion! This method doesn't enable them to race through the sea, but it helps them steer themselves a bit as they drift largely at the mercy of tides and currents.

Some of these aggregates of salps contain thousands of animals, each one up to six inches in length and perhaps two to three inches in diameter. Divers in Australian waters have found a horde of salps measuring more than 100 feet in length and nearly 30 feet across. The divers were able to swim around the mass of animals and observe them closely. They found the commune apparently moving along by jetting in their primitive way in the same direction.

***Salps.*** *These creatures have almost as little substance as the water around them. They propel themselves by muscular expansions and contractions, and frequently live and travel in colonies, giving them the added strength of numbers.*

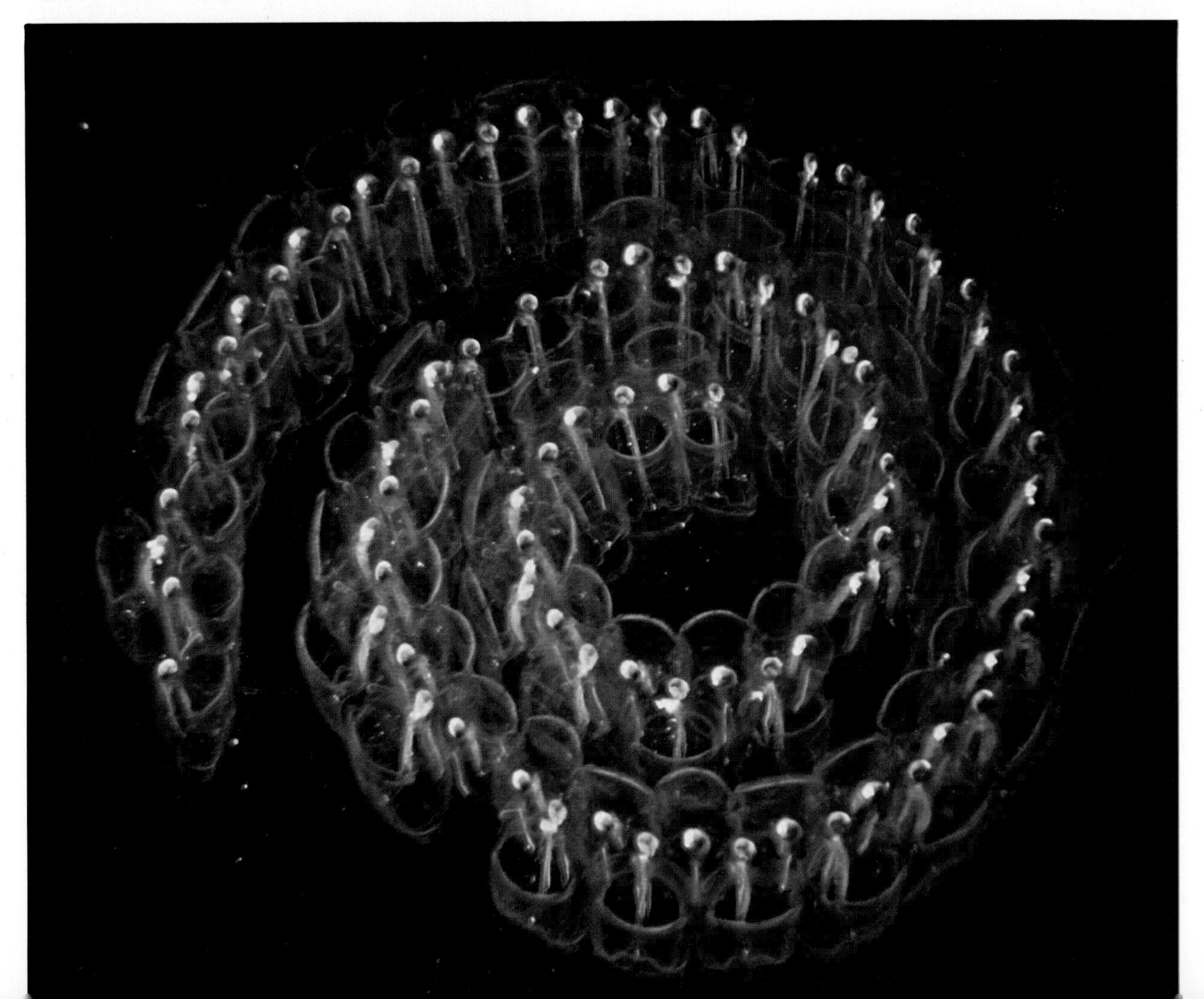

# The Jetting Activity of the Octopus

While pursuing the octopus, our divers have noted a characteristic escape response. Initially when approached, the octopus will freeze and camouflage itself perfectly. It does this by resting on the bottom in a compact mass and assuming the color of the background. As the diver closes in to inspect more closely, the octopus is usually induced to make a run for it. It is at this time that the octopus swims its fastest, using its jet propulsion. Often this movement is accompanied by the release of a puff of ink at each contraction of the mantle. This effort propels the octopus beyond the reach of the diver, but as the diver follows, the octopus usually decides to try the camouflage technique again. This change in tactics may be the result of swimming fatigue—we have often noticed that each subsequent flight is a little shorter than the last.

> "The change in tactics may be the result of swimming fatigue."

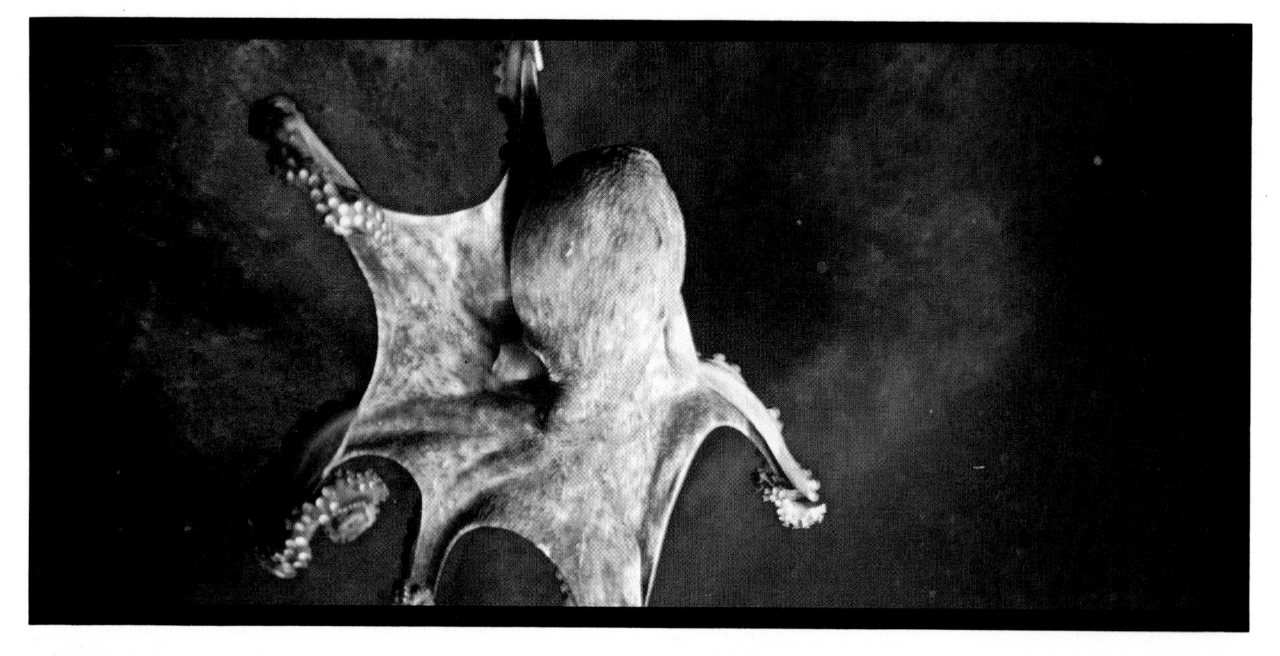

***1 / Orienting itself.*** *An octopus, above, looks around before it descends to the ocean's floor.*

***2 / Jetting.*** *Then, below, it contracts its mantle and expels the water through its siphon.*

***3 / To jet again.*** *Above, it relaxes its mantle and takes in water, then jets the water out again.*

***4 / A gentle jet.*** *The octopus below now jets more gently along the sea floor, seeking a place to hide.*

***5 / Stopping.*** *Below, the octopus extends its arms, stops jetting, and rests on the bottom.*

## A Man-Made Jet-Setter

The diving saucer from the research ship *Calypso* is a member of the jet set. It uses the same type of propulsion utilized by a number of animals. Like the siphon of the octopus, the saucer's nozzles can be pointed in any direction. Jets provide not only propulsion for this unique undersea exploration craft but unique maneuverability and safety, as jet nozzles will not as easily get fouled as propellers.

***Versatility.*** *On the bottom the "minisub" (above) can change direction instantly as it responds to its jet propulsion system.*

***Water jets.*** *The diving saucer (right) blows water from its propulsion jets as it surfaces; this enables its mother ship to locate it more rapidly.*

# Chapter VI. Walkers and Crawlers

It is thought that the coelacanth lives on the sea bottom in moderate depths of 500 to 1500 feet, stalking its prey on its four leg-like fins.

Some marine mammals—sea lions, walruses, and fur seals—also walk, or perhaps gallop is a more apt word, at the edge of the sea. Their limbs are now better adapted for movement through water than for movement over land.

Some animals have several means of locomotion. The octopus, for example, employs jet propulsion, but it can also creep along, using its tentacles with their suction discs. It uses the second method in situations where it doesn't have to move quickly—for example, when it is bringing building material to its home or decorating items for its garden.

The tiniest of animals, the single-celled protozoans, move by any of three methods. Some have minute cilia, or hairlike projections, they use to set up currents that help them along. Others have whiplike appendages with which they flail their way more efficiently. And some move by extending pseudopods, or false feet, ahead of them, then flowing in after the pseudopod.

Among the snails the conch is a leaper. Because it has a narrow muscular foot, it can't just ooze along like so many other gastropods do. Instead it thrusts its foot into the sand to push itself forward. To avoid predators, it can push off and leap perhaps half the length of its shell with every move. Many bivalves, the group that includes clams, oysters, scallops, and mussels, extend a muscular foot and by alternate contractions and expansions creep along.

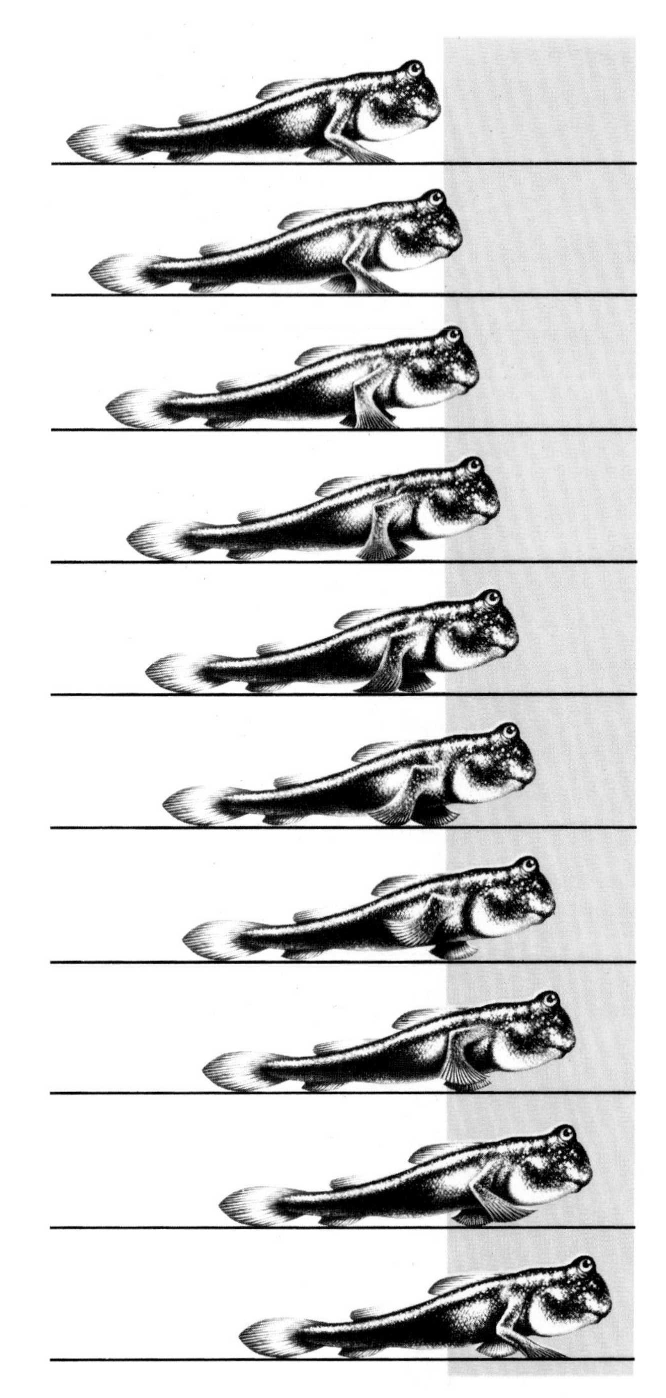

***The fish that walks on land.*** *Above you get an idea of how the mudskipper of the Indian and Pacific oceans cavorts across muddy tidal flats. Out of the water, it raises itself up on its forelimbs and levers itself forward. Then it moves its pelvic fins up and pushes off. Fearless but not foolhardy, mudskippers constantly check their surroundings for danger.*

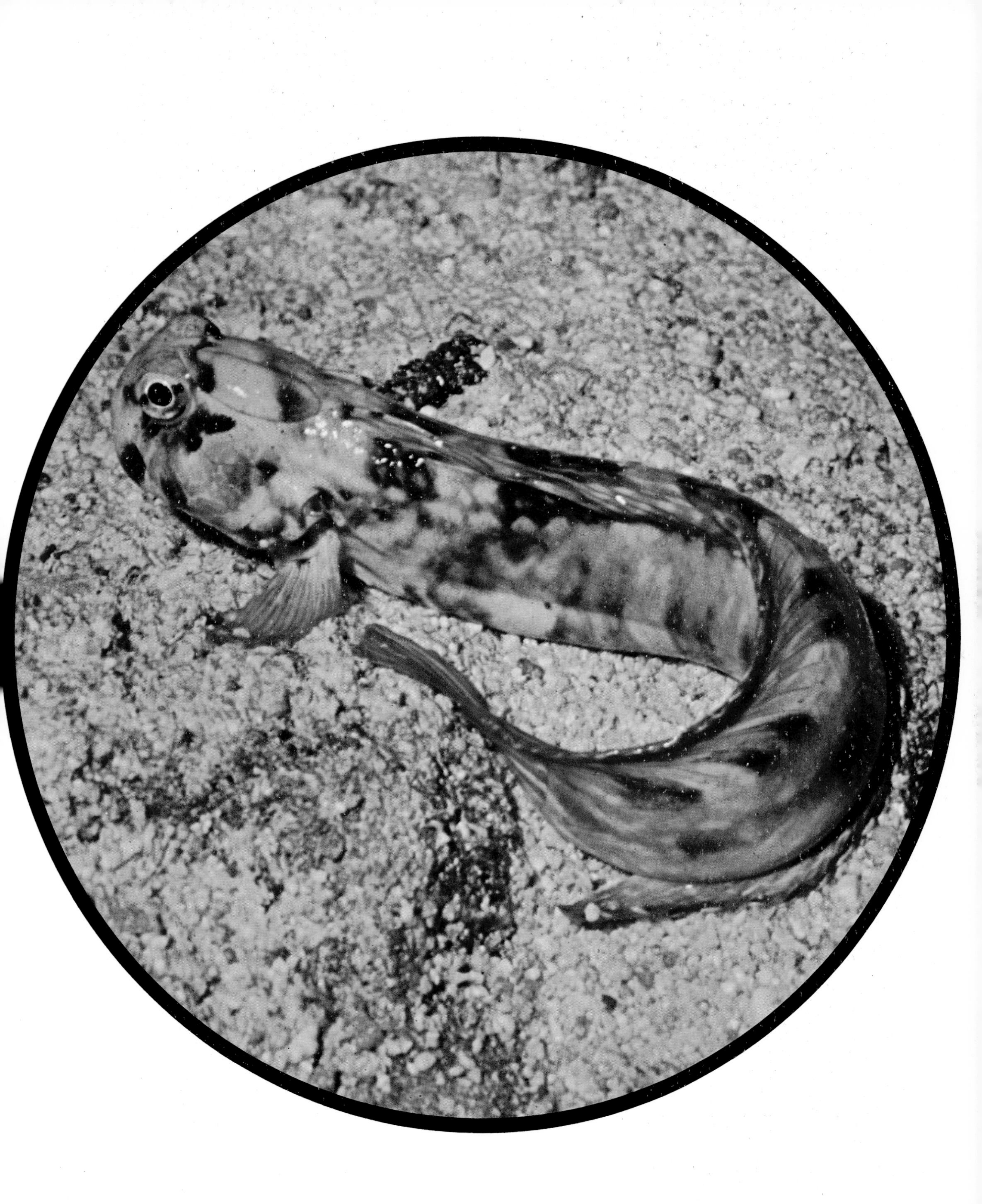

***The macruran shrimp*** *walks and swims about in tide pools at the edge of the sea. As it feeds off carrion in the shallows, it crawls, using its walking legs and its swimmerets.*

## Cautious Stalking

Lobsters, crabs, crayfish and shrimps are animals that wear their skeletons on the outside. Many appendages protrude from these suits of armor, most of them adapted for special activities. Some are used to hold eggs near the animal's body until they are hatched. Others are formed into claws used for fighting, or for capturing and handling food. Four of their five pairs of jointed legs are used to walk about on the ocean floor.

Their armored suits make walking an awkward process for these many-legged crustaceans, for they limit the distance and

***The spiny lobster*** *walks on legs, using swimmerets to help out. These animals spend much of their time in rocky overhangs, but when they come out to feed, they stalk proudly along the sea floor.*

direction their legs can spread. Because many of these crustaceans are scavengers, this method of propulsion is adequate for them to amble across the bottom picking up animal and vegetable detritus. If disturbed, shrimp and lobsters can swim very quickly backward, by flipping the tail underneath in a forward direction.

Members of this ten-legged order (Decapoda) are able to regenerate limbs they have lost. In experiments all ten legs were removed from spiny lobsters. While they regenerated their missing parts, the animals showed a marked craving for calcium carbonate, becoming cannibals if they were deprived of the element.

***The sea robin*** *shown here uses three of its pectoral spiny rays and its body to move about on the sea floor.*

## Walking on Fins . . .

The sea robin is a fish that "walks" on the bottom of the ocean. It walks on the anterior spiny rays of its pectoral fins, which contain taste buds. As it moves and tastes the bottom, the motion of the rays gives the appearance of someone drumming his fingers on a tabletop. The rest of the sea robin's huge, fanlike pectoral fins are spread out parallel to the bottom. It also swings the hindmost part of its body to propel itself forward. As the sea robin strolls along, it frequently stops to dig into the sand with its fingerlike pectoral spines to root out the smaller animals it feeds on. When it isn't traveling on the bottom, it may swim free in midwater, with its pectoral fins tucked tightly in at its side and overlapping its anal fin. When threatened it may dig into sand.

When sea robins use their pectoral spiny rays to grub in mud, sand, or weeds in search of the small animals they feed on, their usual prey are sedentary animals: tiny crustaceans, molluscs, and worms. Occasionally they eat herring, menhaden, and small winter flounder which requires them to move quickly. Anglers have taken sea robins while trolling for mackerel with spinners, indicating they can swim moderately fast when they are not walking the bottom. Although they apparently do not breed in the northern waters, they are not uncommon in those waters, indicating they probably wander there in search of food. So despite their apparent benthic life-style, there is evidence sea robins do a fair amount of near-surface swimming.

## ... And Again

The batfish is another, even more grotesque fish that "walks" the ocean floor. It has no swim bladder, and therefore it tends to stay at the bottom of the sea, in shallows and in deeper water, where it elbows its way. Its pectoral fins are jointed and heavily built. It is on those strong, jointed pectorals that the batfish walks. It uses pelvic fins to a lesser extent in walking. Sometimes it breaks into different gaits—hopping like a rabbit or loping along in the fashion of wild dogs but not nearly as fast. When batfish do get off the bottom and swim, it's with an ungainly, awkward swimming style. They are broad and flat, and they live in even areas that are free of obstructions. Quite often, as a result of their preferred habitat, they are taken by commercial fishermen in their bottom-scraping trawlnets. Their flesh is tender and tasty, but batfish have only a small commercial value, probably because of their ugly appearance. They can be dried out without decomposing, and in the Far East dried specimens are hollowed out. Pebbles are put inside, spines smoothed down and the dried batfish serves as a baby's rattle. They are of interest to the aquarium keeper.

The batfishes number about 60 species, found almost exclusively in tropical oceans of the world. They grow to a maximum size of about 12 inches, and they eat a wide variety of animals, including crabs, smaller fishes, worms, and molluscs, which they capture after lying in wait, hidden by their cryptic coloration and sometimes by covering themselves with sand. Approached by a diver, a batfish will freeze or perhaps cover itself with sand, then lie motionless until the diver leaves.

***This batfish*** *is awkwardly walking on its "elbows" along the ocean floor. The elbows actually are a part of the fish's pectoral fins.*

## Versatility of Movement

*Nudibranchs* (above) *are swimmers some of the time. But they also resort to the slower means of locomotion of crawling, especially when they are feeding. Then they move by the muscular wave action like snails and many other bottom-dwelling marine animals.*

***Whelk.*** *A big muscular foot extends from the shell of the whelk, the large snail pictured at right. And on that foot the snail travels, leaving a trail of slime to grease the material under itself. This whelk carries some unusual passengers—sea cucumbers, which are related to the sea stars.*

A key word for survival in the sea is versatility. This is especially true about the means animals in the sea use for propulsion. Those that are versatile in their modes of movement perhaps have a slightly better chance of survival. By being able to move about in more than one way, animals become a little less predictable. And by being less predictable, they increase their chances of escaping predators. Snails, with ponderous shells, generally move quite slowly, creeping forward on a foot that lays down a carpet of slime. They leave a trail that can be followed by a predator with a sense of smell. But when some species of snails perceive danger they can leap into action to escape. Some somersault their way to safety, thrusting their bodies forward in jumps as long or longer than their shells. Similarly, some clams inch along at an amazingly slow rate until alarmed. Then, they can burrow faster than a clam digger can dig. Worms that crawl slowly on the ocean floor much of the time

"Some snails somersault their way to safety, thrusting their bodies forward in jumps as long or longer than their shells."

can take off and swim very much like small sea snakes at a much greater speed. Some nudibranchs are able to swim by graceful undulations of their bodies that work like all other traveling waves. Much of the time they crawl slowly on the bottom and over obstacles in their constant search for food. There are some other molluscs with wing-like extensions of the lateral foot that are able to swim with a flapping motion of these extensions. Sometimes, the mollusc is the pursuer and can swim to catch its prey.

Sea cucumbers are perhaps a little slower even than their relatives, the poky sea stars. Some of them have tube feet similar to those of the sea stars; others, without tube feet, move by contracting their bodies and by using their anchorlike tentacles to give them traction on the sea floor. Mostly, they move about in their search for food rather than as a means of escape from predators because there are few animals other than man and sea gulls that prey on them. Their tough exteriors help protect them.

## How Various Starfish Move

Sea stars have a very complex, not very efficient, system of propulsion. To understand it, we must look at their anatomy and the equipment they have for moving.

These marine animals, often called starfish, have flattened bodies, and the most common of them have five arms extending from a central disc. There are some species that have only four arms, others have as many as forty. These are not appendages to the body, but part of it; each arm contains branches of the animal's systems. An arm separated from the central disc can regenerate a new body, and a replacement for the arm will be regenerated by the old disc.

The central disc and arms are covered with a skeleton of shell-like plates or rods that are loosely meshed together to allow the animal great flexibility. The sea star's mouth is on the side of the body facing the sea floor. On this same side, along the center of each arm is a V-shaped furrow, called the ambulacral groove. This groove holds nerves, blood vessels, and a water canal, all radiating from the central disc. Outlining the ambulacral groove are rows of tiny muscular tube feet, which end in suction cups in some stars, points in others. These are protected by movable spines.

It is on these thousands of tubes that the sea star moves. But it is not as simple as extending them in the desired direction and then pulling the body forward. The terrain the sea star moves across is often soft, slippery, uneven, or unstable. Unless the star is climbing, say on an aquarium's side, it does not use its suction cups to pull itself forward. Instead, it depends upon the pushing action of its tube feet, and it uses hydraulic pressure

places them firmly, and uses them as levers to push its body along. Some of the feet will extend farther than others to conform to the substrate being traversed. While some are expelling water, others will be filling. All of the sea star's thousands of feet (some are estimated to have up to 40,000 of them) must be coordinated in order to move the animal effectively. It is not surprising that most sea stars are unable to move rapidly. Their average speed is about six inches a minute. A sea star that has been upset is able to right itself in one of two ways. It may pull all of its arms together around the mouth to make its body a tulip shape. Or the star bends its arms away from the mouth until it stands on their tips and topples with some of its feet down.

Having no head or tail, a sea star never needs to turn around. Rudimentary "eyes" and sensitive feelers at the ends of each arm tell the animal the direction it should go. Occasionally the directions can be at variance with one another. One five-armed star was seen attempting to go in all directions. It succeeded in tearing itself into five parts.

***Turning over.*** *In this sequence, reading from left to right, an oreaster sea star rights itself by the so-called somersault method. It uses its tube feet, which are extended, and the suction cups on the ends of them to pull itself around and over.*

***Basket stars*** *(right) spend the daylight hours curled up into an unrecognizable ball. As nighttime approaches, they unravel themselves, revealing five arms, each with numerous branches that they stretch across currents as a large trap for small prey.*

for this thrust. Water that is channeled from the central disc flows into little bulbs above each individual tube foot. By muscular contraction of the bulb, water is forced into the tube through a nonreturn valve and extends the tube foot. The tube is shortened by expelling the water through the pressure of the tube's longitudinal muscles and relaxation of those of the bulb.

When a sea star wishes to move, it extends some of its tube feet in the desired direction,

***Sea urchins,*** *armed with sharp spines, frequently graze on algae as they proceed along the sea floor at a barely discernible speed. It requires time-lapse photography to see the patterns of their movements.*

## Slow Movers

Sea urchins and sea cucumbers are relatives of the sea stars and generally move as slowly. Sea urchins have tube feet as sea stars do, only those of sea urchins are more slender and longer, reaching out beyond their spines. They move by means of their tube feet or their spines or a combination of both. The oral surface, that is, the side that rests on the sea floor, has the tube feet and spines that are used to move about. The tube feet, like those of some sea stars, have powerful suction discs at their ends, enabling sea urchins

***This sea cucumber*** *appears to be lying quietly on a sandy bottom. A few sea cucumbers just don't bother to move around in search of food—they simply let ocean currents carry it to them.*

to climb slick vertical surfaces. Using their tube feet, some sea urchins have been clocked at speeds averaging about an inch a minute or five feet an hour. Others have been clocked as fast as six inches a minute. Traveling on the tips of their spines, one species of sea urchins has been clocked at speeds up to five feet a minute—or 300 feet an hour. This species travels in groups in what appear to be extensive migrations. Some sea cucumbers have tube feet which they use for locomotion. These and others without tube feet also use wormlike contractions of their body muscles to move.

## Trails on the Seafloor

When man first lowered cameras into the depths of the sea and took pictures of the abyssal ocean floor, he found evidence that someone or something had been there before. All over the sea floor were markings of some kind which proved, on close examination of the photographs, to be animal tracks. More cameras with lights were sent down and more pictures were taken and more tracks were discovered. Eventually, man was able to get down to even the greatest depths personally to inspect this mysterious ocean bottom and study the animal tracks firsthand instead of from photographic evidence. What he discovered deep in the abyssal portions of the sea were not only animal tracks but the animals that made those tracks, and evidence of an intense underground life deep within the bottom sediments. At last there was eyewitness, indisputable evidence that there is, in fact, life in the greatest depths of the world's oceans.

*__Animals on ocean floor.__ Above we can see short-spined echinoderms called sea biscuits, pushing their way along the soft-bottomed ocean floor. They are related to the nearby sea urchins and sand dollars.*

What did the scientists find when they investigated the great depths of the sea? What did they see crawling around on the sea floor, making those curious little tracks and trails? They saw a collection of some of the strangest animals man has yet found. And they also found some very ordinary looking creatures that might just as easily have been living on the continental shelf.

Among the curiosities they found the blind tripodfish, *Bathypterois,* standing on the bottom on a couple of rays from its two pectoral fins and on a third ray from the lower lobe of its caudal fin. There were deep-sea codfish, sharks snuggling into hollows on the bottom, and sea urchins laboriously moving over the sea floor. There were sea stars and brittle stars and crinoids. They found crustaceans and several kinds of seaworms.

# Chapter VII. Other Ways to Go

The quickest animals that live in the sea get around by swimming, beating their fins, undulating their bodies in a traveling wave, or using jet propulsion. Slower creatures walk or crawl on the bottom. Still others stay in one place, taking sustenance from the passing waters.

For centuries man depended on the power of winds or currents to take his vessels across the seas. Finally, by trial and error, a totally submerged propeller with two or more blades was developed. With refinements, it is the motive mechanism found on most ships today. It is tailored to suit the specific vessel it serves, whether it is an atomic submarine, a passenger liner, an oil tanker, or the family launch.

> "Some dinoflagellates travel 150 feet each day. This sounds like a short distance to us—but the equivalent for man would be a daily 2000-mile underwater journey."

But this is a costly answer to water travel and gives man little prospect for improvement. We look again to other animals for clues. Dinoflagellates, tiny plant-animals, have two whiplike organs they beat against the water to propel themselves, pushing the water behind them. Some dinoflagellates travel 150 feet each day in vertical migrations. This sounds like a very short distance to us, but it is almost 2 million times the length of the dinoflagellate. The equivalent for man would be a daily 2000-mile underwater journey, clearly beyond even our most advanced technical capabilities.

Looking to animals more similar to man, we compare ourselves with other mammals that have returned to the sea and find many similarities in technique. Like an experienced aqualunger, who uses his arms in swimming mainly to turn, a sea mammal with armlike appendages usually holds them close to its body. The turtle's use of its flippers as paddles for swimming is not unlike the underwater swimmer using his arms to do the breast stroke. Nor is the penguin's use of its wings for underwater flying much different.

We have copied the webbed feet of seabirds and the flippers of seals for swimming. Flippers give us more surface to push against the water, but our legs are peculiarly unsuited for an aquatic life. Many serious swimmers, interested in improving their technique, have attempted to emulate the dolphin's swimming movements. To do this, they keep their legs together in one unit similar to the dolphin's flukes and flex their legs from the hips with only a suggestion of knee action. Swimming this way, a man with only one leg can compete effectively with a man who has two. But we are still left far behind in the wake of the dolphin.

***Electric motor submarines.*** *With the advent of the aqualung, people could go underwater free of surface connections. But to move about beneath the surface with a minimum of effort and thus conserve air, divers began designing and building wet submarines. The most primitive were towed by surface craft, defeating the purpose of being free of topside support. A step along the road to independence was the pedal-powered submarine, but this forced the diver to expend energy and thus use up his air supply more quickly. The most advanced of these wet submarines are powered by electric motors much as military submarines. Some carry two aqualung divers; some carry one as pictured here. In this submarine the electric motor is sealed in the stern; it is connected directly to a reduction gear, also sealed watertight, and that, in turn, is connected to the propeller, which provides the needed drive.*

## Man's Undersea Limitations

Basically the size and weight of a porpoise, a human being also displays a comparable anatomy. But the analogy ends there. The porpoise's physiology allows him to stay underwater without breathing ten times longer than man; to dive five times deeper; to swim at least ten times faster; to see clearly above and below water, while man is practically blind under the sea; and to be practically insensitive to water temperature, while man survives but a few minutes in ice-cold seas. We will see in another volume what technology can do (including surgery) to improve human physiology under the sea. But a still formidable obstacle is self propulsion, for which we are basically crippled. The ocean is vast in three dimensions; and swimming unaided, even the most sophisticated *Homo aquaticus* can only be master of a very small piece of territory.

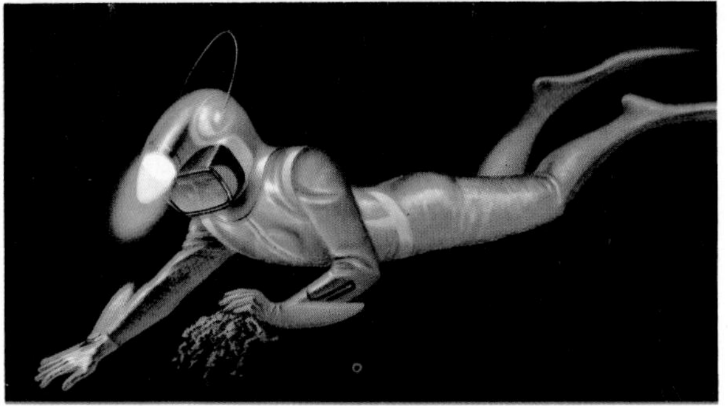

**Homo aquaticus.** *His lungs will be bypassed and his blood supplied with oxygen from cartridges.*

***Man's swimming capabilities.*** *Compared to sea animals, a human moves through the water with a very low rate of efficiency. The addition of swimming equipment helps considerably, but it still doesn't greatly increase our speed or efficiency. Without equipment, the fastest way to move is to plane along on the surface. Surfers know this and use a flat board to present a plane gliding surface to the water. (They add a skeg, or keellike device, on the bottom of the board to give some small measure of stability.) We can do away with the board, of course, and body-surf, as this swimmer is doing. Or we can flail our arms and legs in more or less regular and rhythmic fashion the way a swimmer does and move through still waters.*

***Iguana.*** *The marine iguana is a mediocre swimmer, using its body and long tail to undulate through the water. It rarely goes more than 100 feet offshore, however, since it finds seaweed close to the coastline.*

## The Waggler

The marine iguana tucks its four legs close to its sides and uses its body to wend its way through the sea. While most lizards are terrestrial, the marine iguana of the Galápagos Islands in the Pacific lives in and at the edge of the ocean. Some of the time it lies quietly in the sun along the rocky shore. Some of the time it swims offshore and dives to the beds of algae on which it feeds. To move through the water, it uses the time-honored method of its relatives: the traveling wave. Through evolution, the tail of the marine iguana has flattened laterally in about the same manner as the tail of the sea snake.

***Walrus.*** *This mammoth animal spends most of its life in the sea, and it is an excellent swimmer. A bull walrus can weigh close to 4000 pounds, and its thick layer of blubber protects it from the icy water.*

## Blubbery Athlete

This huge, ungainly, blubbery animal, which often wallows on the ice floes of the Arctic Ocean, becomes a graceful, smooth, efficient swimmer when it slips slickly into the sea. The transformation is astonishing because the contrast is so great. On land the walrus moves so ponderously that an observer feels sorry for it. But the same observer must feel a sense of wonder when he sees how graceful the enormous beast becomes in its own element. The walrus tucks its forelimbs out of the way when it oozes itself into the water and uses its hindlegs—broad, flattened, paddlelike limbs—to propel itself.

***Adélie penguin.*** *This is one of the species of penguins which live on Antarctica. It does not fly, but uses its wings to swim.*

***Double-crested cormorant.*** *Large and powerful, with webbed toes and a hooked beak, this bird is a voracious eater and excellent fisher.*

## Webbed-Foot Swimmers

There are fifteen species of penguins, found in various parts of the Southern Hemisphere and as far north as the Galápagos Islands. Seven of these species live almost exclusively in the frigid waters of the antarctic, which are rich in their basic food, shrimp named "krill." Penguins spend most of their lives in the sea; they "fly" very swiftly underwater, using powerful strokes of their short, smooth wings. Their webbed feet are used mainly as rudders. Some species are capable of cruising great distances at seven knots and to dive as deep as 900 feet. Cormorants also "fly" underwater with their big wings and can cover at least half a mile without surfacing.

***Manta ray.*** *Also known as the "devil ray," these animals can weigh as much as 3000 pounds. The manta ray makes beautiful leaps out of the water.*

## Underwater Fliers

The manta ray is another underwater flier. There is a certain similarity in the underwater swimming of birds like penguins and cormorants to that of fish like mantas, skates, and other rays. In each case the forelimbs, analogous with our arms, are used for underwater flight. While birds use their wings, mantas use their pectoral fins, which are developed into enormous, flexible, triangular, winglike structures. And by flapping they change the pitch of the wing to get the utmost efficiency during upbeats as well as downbeats. The result is a graceful ballet. The manta's reputation as a beautiful swimmer is beginning to surpass its undeserved reputation as a devilfish.

## Paddlers

Man's racing oars have flat blades, slightly curved and of the right length-to-width proportion for top efficiency. They're lightweight to enable the oarsmen to swing them easily to the coxswain's call, sometimes as many as 40 strokes a minute in hard-fought competition. The oarsmen's seats slide back and forth in response to their strokes. The most efficient undersea paddlers are the giant sea turtles. Clumsy, almost helpless when they have to creep ashore to lay their eggs, they are fast migrators in the sea, can reach peak speeds of ten knots, and are very agile.

***Sea turtles,*** *like the hawksbill above, are so completely aquatic in life-style that only the females ever leave the water and then only to lay eggs on the beach. Once the eggs hatch, the hatchlings immediately head for the sea.*

***The fastest man-powered boat.*** *The most efficient man-powered watercraft is very likely this eight-oared racing shell (right). Working as a closely coordinated team, the oarsmen each swing one oar in unison with all the others.*

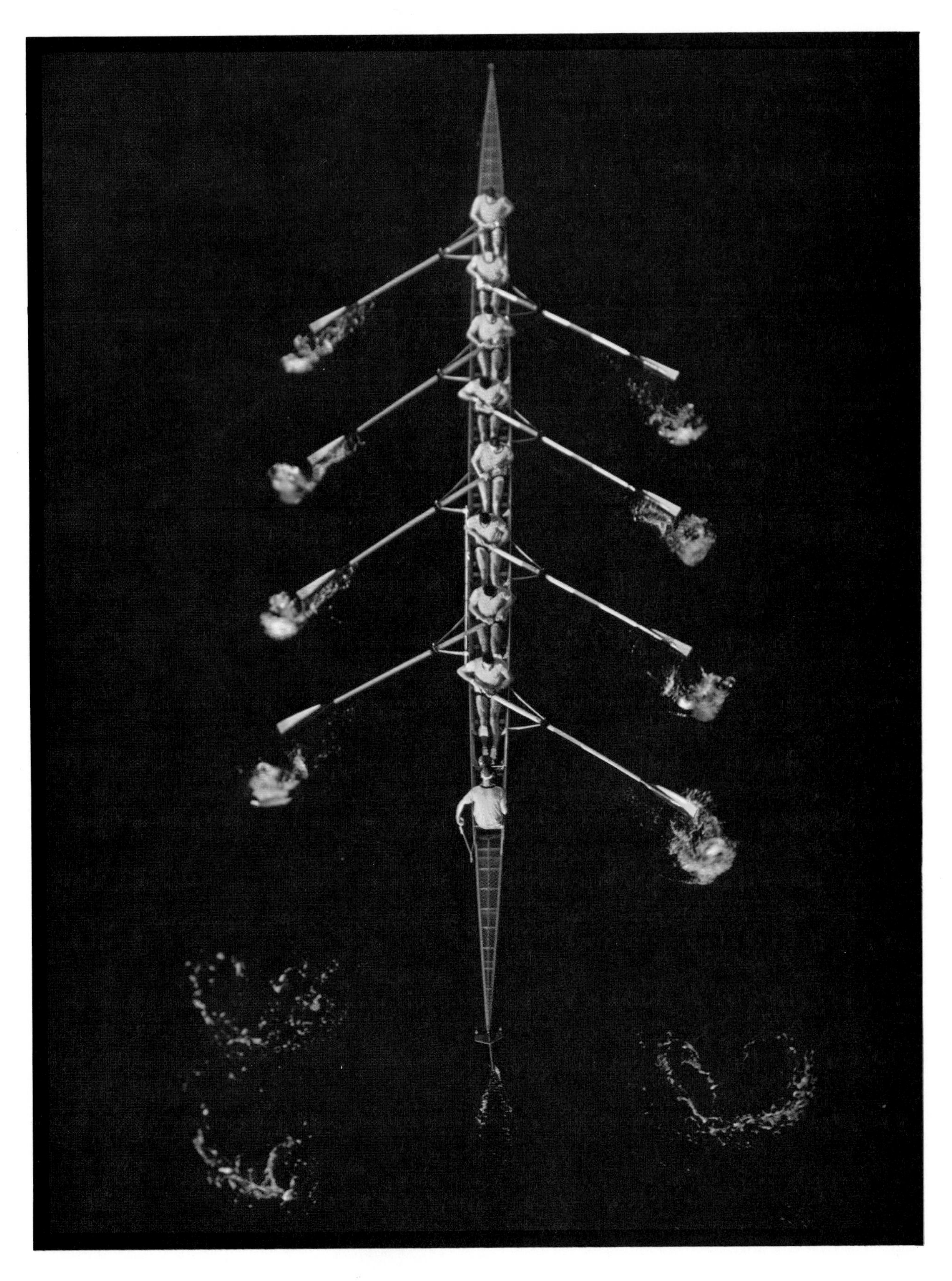

## Catching the Wind

By twisting its sail, a tiny velella is able to steer itself slightly, but practically it is at the mercy of the winds and is often stranded on beaches. Beneath its float of air cells, there are many short tentacles with which the by-the-wind sailor stings and captures minute living creatures and passes them to its mouth, which is centered on its underside. Great flotillas of these little animals are frequently seen in tropical waters, dotting the ocean for miles in all directions. Fortunately, their sting is harmless to man.

Single triangular sails have plied the coastal waters of the Indian Ocean and the Red Sea since the ancient Egyptian and Phoenician sailors of 5000 years ago. Square sails were common until less than 100 years ago when the more efficient triangular sails became widespread. Until they were supplanted by steam and diesel engines, sails powered all the world's ships. Modern yachtsmen have a background in aero- and hydrodynamics; they are able to tack pretty close to head winds. They have learned also that certain winds can be counted on at certain times of year to help in ocean crossings and competitions. Racing sailboats have become very sophisticated.

***A living sail.*** *Usually the tiny iridescent sail of velella, the by-the-wind sailor (above), stands high in the breeze that pushes the little hydrozoan across the surface of the sea.*

***Man-made sails.*** *Like the velella, man has harnessed the wind for many centuries, using it to send him across the surface of the sea in boats like these Arab dhows pictured at right.*

# Chapter VIII. Hitchhikers

An attractive way of life is that of the marine hitchhiker. It is a rare event to find any of the sea's large and powerful creatures without at least a few freeloaders accompanying it. Whales, dolphins, manta rays, turtles, man's vessels, and many others play host to one or several easy riders. The hitchhikers' mischief ranges from the harmless capering of the little pilotfish, swimming effortlessly in the compression wave at the snout of a shark, to the deadly attachment of the sea lamprey, who is not so much interested in transportation as in its host's blood.

> "The hitchhikers' mischief ranges from the harmless capering of the little pilotfish to the deadly attachment of the sea lamprey."

There are a number of advantages to be gained by hitching a ride. If one cannot swim, as in the case of the barnacle, or if changing location is difficult, as it is for the sea anemones, the host can take his guest to places where food will be abundant. Species that have a wide distribution throughout the seas have a far better chance of survival than those concentrated in one locale. Then no localized phenomenon—a change in the water's temperature or chemical makeup—will be able to wipe out the whole population.

The distance a hermit crab takes its passenger, the sea anemone, is not the most important factor in this driver-rider relationship. These crabs live in the abandoned shells of other animals and occasionally must relocate to larger ones. The crabs appear to want to carry anemones along as decoration, and they probably gain some protection from the anemone's stinging tentacles. At any rate, the association has been observed very frequently; the anemone's tentacles are often positioned close to the crab's mouth, thus enabling the rider to catch a few crumbs from the host's table.

Other hitchhikers return no favors to their benefactors. They are along on the ride for what they can get.

Flatworms, roundworms, tapeworms, and threadworms invade the intestines, heart, liver, muscles, and bloodstream of almost all species of fish. In addition to these internal parasites, many animals are afflicted with fish lice, a crustaceanlike animal that fixes itself with a circular sucker to the outside of a fish and changes its position at will. A slightly different parasite attacks fish in the same way as do fish lice.

Disagreeable as they are, most parasites allow their victims to live. Not so the sea lamprey. With a disklike, rasping mouth, the lamprey attaches itself to its victim, drills a hole in it, and sucks its blood. When there is nothing more to gain from the unfortunate host, the lamprey drops off and sets out to find another ride.

***Clusters of whale lice** have settled on this right whale's head and body, and they get a free ride wherever the whale carries them. These little shrimplike creatures hook their legs into the skin of the host and gnaw out pits in which they live protected from the wash of water passing over the whale as it swims. The whale louse takes all of its nourishment from its host. Whales that pass through warmer water than does the right whale also play host to a few species of barnacles. These feed on phytoplankton, the microscopic drifting plants of the sea, by kicking it into their mouths with their feet. Attached to a constantly moving whale, the barnacles are carried through unharvested waters, which are often rich in the nutrients that sustain barnacles.*

**Red wrasse host.** *Above, playing host to an isopod, this wrasse is deriving no benefit from its hitchhiker. The parasitic isopod is deriving sustenance from its traveling companion.*

## Parasitic Crustacea and Fish

Isopods are tiny, sometimes microscopic crustaceans that inhabit the waters of the world. There are some that have left their benthic habitat to ride the coastal waters aboard a finned host. Most isopods live on the bottom under rocks or camouflaged among algae and sponges. One group of isopods, commonly known as gribbles, lives in pilings eating wood like a termite and may travel only a few inches in a lifetime. They are all relatives of the common sow or pill bug. The parasitic isopods are travelers, although not by their own means. These are crustaceans which attach themselves tenaciously to the skin of a fish and feed off its life blood. Others grasp the gills or some other part of the luckless fish that must support it. Some species of parasitic crustacea or fish lice, in fact, are choosy about where they attach themselves to their unwilling hosts, selecting a specific part of the body. To attach themselves in a manner that makes them almost unshakable, these crustacea use six or seven pairs of sharp penetrating hooks.

The shape of such a parasite probably does not affect the swimming efficiency of its host appreciably but its size may. Now and then a young fish is seen with a large isopod attached. Such an isopod may create a wound open to infection. It is the secondary infection of bacteria or fungus that may harm the host more than the actual isopod. When a host dies, the uncaring parasite leaves to find another one to support it.

## Remoras and Shark

Remoras or suckerfish are among the strangest hitchhikers in the sea. For they benefit in the commensal relationship they enjoy with other sea animals, and their companion is unaffected. Stranger still is the manner in which a remora attaches itself to its commensal. Atop the head of the remora is an oval shaped suction grip device that looks very much like the grill of an automobile or the grating on an air conditioner. By moving the transverse parts within the oval, the remora can attach itself firmly to any surface. By moving those parts in another way, it can release itself. But this attachment process has no effect on the commensal. It doesn't even leave a mark. The strangest part of all this probably is that the oval-shaped organ that the remora attaches with is really its dorsal fin, which has migrated forward to a position atop the head and changed in form and function. In other fish the main purpose of the dorsal fin is lateral stability. Not so in the remora's case. It is solely for attaching itself to a commensal. Remoras frequently attach to large animals other than sharks. Groupers—especially the big 500 or 1000 pounders—sea turtles, whales, and occasionally a human diver have attracted remoras and carried them about. Mainly, the remora gets a free ride out of this relationship. But in addition, it is there when the commensal partner is feeding. If that partner should drop a few scraps of food, the remora is quick to detach, snatch a morsel, and reattach. It is also believed to clean its big host of parasites.

***Shark with chinfish.*** *In the photograph below a whale shark is seen with many remoras attached to its chin. This relationship is called commensalism—the remoras are getting a free ride and some food, and the shark is not being harmed.*

## Dolphins Riding Bow Wave

As a ship plows its way through the ocean, it creates waves. Occasionally dolphins come to play and ride on these waves. Sailors at sea have often seen these playful mammals riding the bow wave of their ships with effortless ease. Dolphins somehow have learned how to get a free ride. From all indications they are not seeking food nor are they being lazy. They probably simply enjoy the game of surf-riding.

One theory of how this phenomenon works has been described by biophysicists. As a ship plows along, water piles up in front of the ship's bow and is forced forward. The waves that are thus formed move with the ship. Dolphins typically approach a ship and scout around the edges of this bow wave area, perhaps five to ten yards from it. This indicates that the dolphins may be "feeling out the pressure field."

Once they have ascertained the conformation of this pressure field, they can either

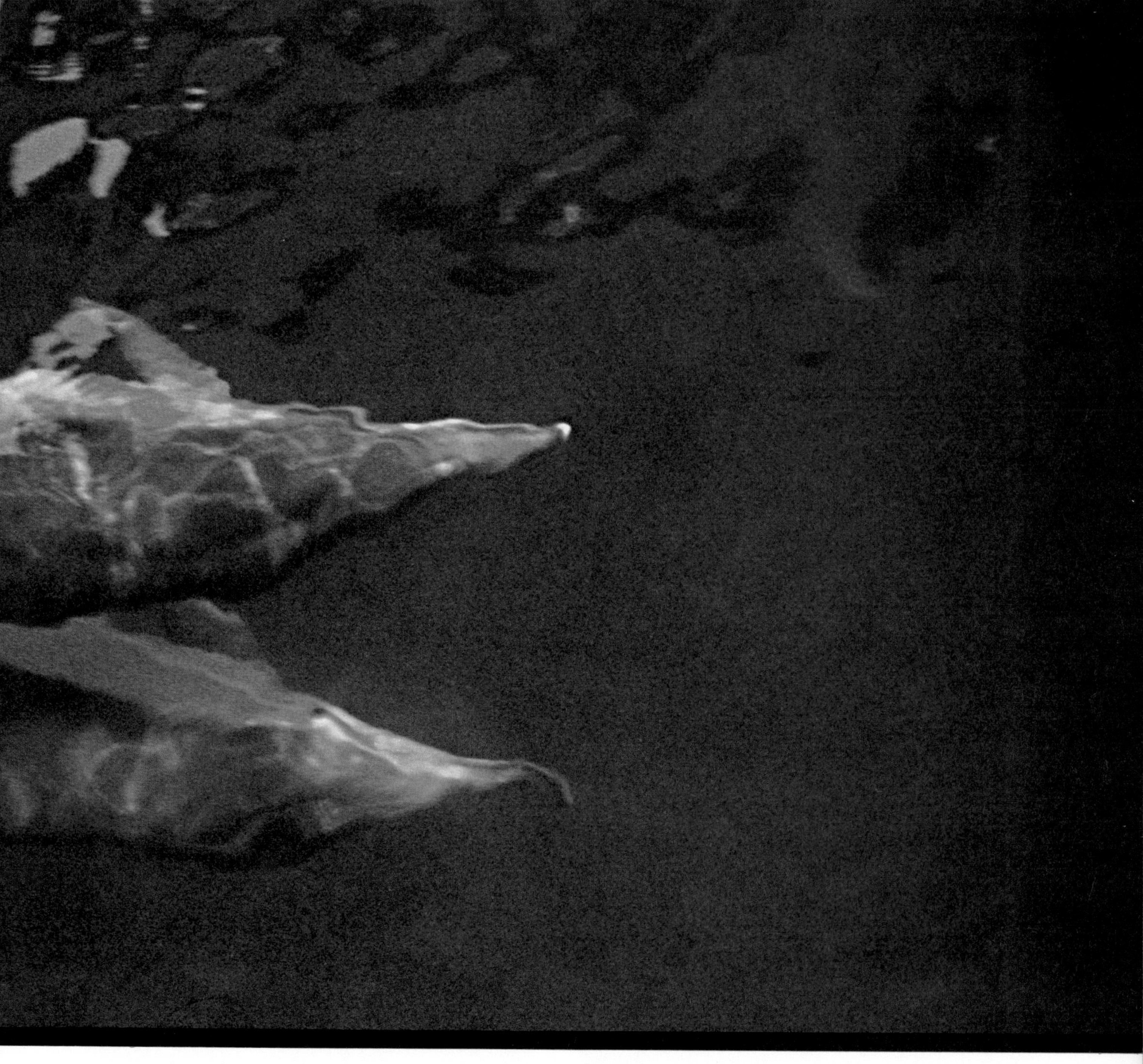

swerve into it or swim away to play elsewhere. If they do move into the bow wave, they position themselves on its forward slope. Gravity keeps them moving downward along the front slope of the wave. And the position in which they hold their flukes gives an additional upward thrust.

***The bow wave*** *of a ship provides a free ride for this pair of Atlantic dolphins. When they are not hitchhiking, dolphins are capable of outdistancing the fastest ocean liners.*

As they ride alongside of ships at sea, dolphins give the appearance of leading the craft toward some safe haven. This impression is at the origin of the legend of the colonization of Crete many centuries B.C. Sailors have always looked upon dolphins as harbingers of a safe voyage. The very playfulness of the animals is another factor that leaves observers with a feeling of goodwill about dolphins.

# Chapter IX. Getting the Hull Out of Water

The sea is a strange environment—sometimes it seems friendly, sometimes not-so-friendly, and sometimes downright malevolent. A denizen of the deep may be wrapped securely in a watery blanket one moment and held in bondage before an onrushing predator the next, unable to flee quickly enough. Different inhabitants of the oceans have adapted to this changeable way of life in different ways. One of the most spectacular methods of escape is to get completely clear of the water, even if for just a few seconds. To illustrate—a member of a scientific expedition in the Gulf of Panama was swimming off an island with a sheer rock cliff for a shoreline, when he was approached by a shark. As the shark angled toward him, perhaps only out of curiosity but perhaps with mayhem in mind, the diving scientist, unused to being explored by sharks, tried to scale the sheer rock cliff. He cleared the water completely before flopping back in with a loud splash. As he looked quickly about him, he saw the shark was no longer in sight, and he was able to get back to the boat from which he had been working. Even that brief absence of a few seconds from the water was enough to send the prowling shark off looking for something else.

Survival isn't the only motive for getting clear of the water. Some animals quite literally jump for joy. Frolicking in the briny is a pleasurable pastime for them. Some other sea creatures leap in anger, fear, or hunger. And there are a myriad of mysterious reasons for jumping, reasons scientists still seek. Whatever it is, the ability to get clear requires special mechanisms as well as special motives. To break free of the sea, a marine animal, whatever its weight, must reach a certain vertical speed. For example, to leap two yards high above the surface, a creature must break the surface with a vertical speed of twelve knots, whether it is a sardine or a blue whale. To reach eight yards, the speed must be twenty-four knots! Consequently, only fast swimmers, large or small, have the privilege of jumping into the atmosphere.

"Some animals quite literally jump for joy. Frolicking in the briny is a pleasurable pastime for them."

The ponderous cowfish is hardly likely to leap clear of the water even with a healthy riptide behind it. It is just not adequate for a venture into the air. So equipment, favorable conditions, and motive are all factors to be combined if a member of the undersea world is ever to make brief intrusions out of its element.

For a variety of reasons men have also sought to get their bodies and their vessels out of the water. Some have acted to escape danger; some to express aggression; some to develop improved transportation; and some just to have fun. Engineers have designed all manner of devices to get us up and out of, as well as down and in, the water.

***A pair of mute swans,*** *after taxiing down the watery runway of their home pond, lift off and get airborne. They use a tremendous amount of energy and must paddle along the surface with their large, webbed feet before they can break free of the water to fly at speeds up to 45 miles an hour. Their speed through the air is far greater than their speed swimming through the water, although their feet are well adapted for surface swimming. Being able to fly gives the swans greater range than if they were restricted to their graceful but slower swimming or their awkward waddling gait over land. Some ducks, with a tremendous burst of energy, can leap directly into the air from a floating start in the water. But most ducks, geese, swans, and other water birds must paddle along the surface to become airborne.*

## Fish That Fly

The adaptation that has enabled the flyingfish to survive through millions of years of predation is its ability to get up and out of the water for extended periods of time—long enough to escape predators. In this case an extended period of time may last up to 20 seconds. As this adaptation continues to develop, it may someday in the far-distant future enable flyingfish to soar through the air above the water for minutes at a time. One flyingfish was seen to reach a height of 36 feet above the sea's surface. And 20-foot-high flights are not uncommon. Soaring along at speeds up to 35 miles an hour, flyingfish may travel 250 yards or more in a single glide. However, flights of 60 to 70 yards, lasting six seconds, are more typical, but can be repeated several times in a row with a short dip of the tail.

Typically, a flyingfish launches itself through the surface first by gathering speed swimming upward toward the surface. This speed helps it leave the water. With the enlarged lower lobe of its caudal fin trailing in the water, it moves it rapidly from side to side and extends its broad winglike pectoral fins. The forward thrust given by the caudal fin plus the lift provided by air acting on the pectorals help the fish become airborne. When it drops back to the surface, it is the whirring action of the caudal fin in the water that helps the fish repeat its flight. Often the predator follows the flyingfish, seeing its silhouette through the surface, waiting for it to splash back. In this case, the flyingfish tries to fool its chaser by making a right angle turn in flight. Some fish, notably the dorado, have been seen to chase flyingfish right out of the water, sometimes capturing them in their mouths in midair.

***A takeoff run*** *is being made by this flyingfish, much as man-made aircraft do. It starts building up speed as it flashes upward toward the sea's surface.*

***Clearing the water.*** *The lower lobe of its caudal fin remains in the water for a moment, whirring like a propeller.*

***Emerging,*** *its extended pectoral fins act like wings do on an airplane and give a certain amount of lift. Its pelvic fins help a little too.*

***Soaring.*** *As the fish gains speed, it expands the pelvic fins, providing lift to raise the tail above the water. It then soars off.*

## Difficult Journey

After they have spent three or four years in the open ocean, feeding and gaining power, the Pacific salmon undertake their final journey to the stream where they were born. Obeying a mysterious instinct, these magnificent swimmers let nothing deter them from their destination—their mating grounds. There salmon eggs will have a good chance to hatch, and the fry will have abundant food to sustain them until they are large enough to tackle the trip seaward. The returning adults travel through rushing rivers at a rate of about five miles an hour.

Rapids and falls are encountered frequently. Then the salmon must leap—spectacularly —to heights of 8 to 10 feet. For some salmon the challenge is too much, they literally batter themselves to death. We know that to leap a waterfall 10 feet high, the salmon must have reached a *vertical* speed of 15 knots. As the rivers are generally very shallow they can only meet this challenge by speeding horizontally, just below the surface, and suddenly steering their momentum upward by a tremendous effort of their pectoral fins, or by rolling on one side just before jumping, bending sharply and forcing themselves in the air by a last stroke.

***Leaping over a cascade.*** *When the impetuous waters of a mountain river tumble over rocks in rapids or cascades, eddies develop in the pool below. In these turbulent areas the direction of the flow is constantly changing, sometimes surging toward the surface and sometimes even back upstream. If the salmon can get into one of these eddies, it can take advantage of the backwash there to get the boost it needs for spectacular leaps.*

***Swimming up a fall.*** *The downstream flow of water in a river is sometimes broken by waterfalls or cascades. If the slope of a rapid is not too steep, or if a fall is not too high, the salmon powerfully swim up the fall. When the fish reaches the top, its enthusiastic swimming sometimes carries it right out of the water. Then, after a momentary tailstand, it plops back into the water and continues its arduous journey.*

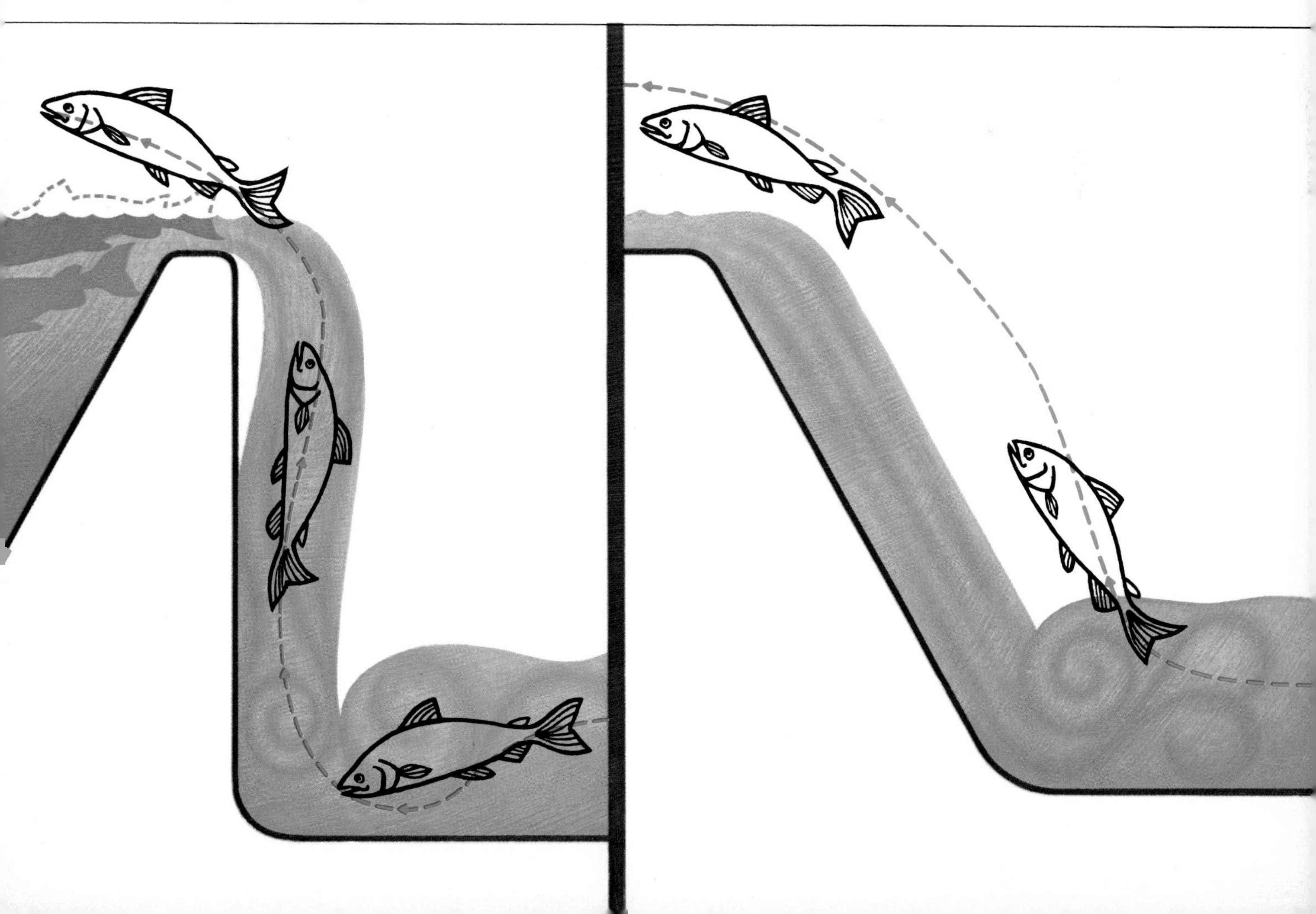

## Jumping Billfish

Swordfish, marlin, and sailfish, collectively known as billfish, are some of the ill-named "game fish." When it has been hooked by a so-called sports angler, the billfish leaps spectacularly out of the water as it attempts to shake loose the barbed and sharpened hook that pierces its mouth. It is no game for the game fish. The angler must keep a taut line to prevent the fish from escaping its intended fate as a futile trophy and photographic model at dockside.

Sailfish, with their high dorsal fins, and marlin, which share a lunate tail with their fellow billfish, are fast, powerful, and they range the open seas of the world. Both have a reinforcing ridge, called a careen, on their caudal peduncle—the narrow part of the body at the base of the tail—which separates the water smoothly as it oscillates back and forth. Some remarkable footage of a swordfish underwater has shown that they do not swim in the same manner as a tuna. Instead of maintaining a relatively rigid body and a rapidly fluttering tail, this swordfish moved through the water in a sinuous manner, not unlike the propulsive mode of an eel or sea snake. The footage showed that, at least at slow speeds, the swordfish utilized most of its body surface for propulsion by passing waves of undulation from its head to its tail, in a basically anguilliforn manner.

***Out of the water.*** *Above, a marlin, perhaps weighing close to 1500 pounds, jumps with the greatest of ease when trying to dislodge a hook.*

***Sailfish close-up.*** *This model (at right) depicts the wonderful leaps of a sailfish in the open sea. Its long bill not only is an effective weapon but may ease water resistance.*

***Mullet.*** *These leaping fish appear to have boundless energy. They jump to escape being caught by man and their fellow sea-dwellers.*

## Leaping for Survival

Mullet often leap clear of the water as they seek to escape danger. The danger from predators may be caused by men seining, as in this picture, or by hungry dolphins or fish. By getting out of the water for even a brief second or two, the mullet may be able to elude the peril that stalks it. Often mullet may be seen leaping by the hundreds from a seemingly placid sea. But somewhere beneath them some danger lurks, probably already making a dash into the midst of the school and decimating its numbers. The position of its pectoral fins—high on the sides of the body just behind its head—helps the mullet fling itself clear of the sea. Schooling also helps the mullet in its fight for survival through safety in numbers. Plus the quickness of its movements.

***Sea lions.*** *These strong swimming mammals move swiftly with powerful strokes of their long front limbs, not unlike the underwater flight of penguins.*

## Leaping for Play

Sea lions cavorting in shallows or playing in deeper seas often leap from the water. Their streamlined shape and muscular bodies with powerful hindlimbs and flippers enable them to do this. Sometimes their leaps are conventional arcs through the air with a headfirst reentry. Other times they may do backflips like this one. Typically, a sea lion will pick up speed underwater and then break through the surface, arching upward and forward through the air before reentering the water. They may leap clear of the water when they chase fish as well as when they are courting or amusing themselves. Their playfulness and their sleek appearance make them favorites of people along the California coast, where they are frequently seen gamboling off the rocky shoreline.

## Whales Clearing the Surface

Whales, like people, often like to vary their pace. Just as people will sometimes take to running or jumping or hopping or skipping, so do whales apparently enjoy lolling about on the surface, jumping high into the air, swimming on one side or the other or even on their backs. Of these variations of movement by whales, jumping, or breaching as it is sometimes called, is the most spectacular. Whales come to the surface in a variety of ways and for at least two reasons. Being air breathers, they must come to the surface for air. If they are lazing along at or near the surface, they allow themselves to rise vertically until they break surface. Quickly, they will exhale the breath they have been holding, take in a new breath of air, close their blowholes and submerge quietly. Or, if they have been moving along underwater at a good rate of speed, they surface at an angle —about 30°—exhale, inhale, close off, and dive back in, all in the space of a few seconds.

***A gray whale spy hops*** *as it checks the Baja California coastline near its breeding ground in Scammon's Lagoon. The bay where gray whales breed is protected water under Mexican law.*

There are other times, though, when whales seem to enjoy leaping high into the air, flopping back into the sea with tremendous resounding smacks against the water's surface.

Humpback whales especially seem to enjoy this sport, slapping the water with their flukes. Sperm whales also jump clear of the water only to fall back in, raising gigantic columns of water. Such spectacular jumps are probably the conclusion of very deep dives, when the whales, out of breath, speed up from 3000 feet for air at more than 20 knots and leap.

The gray whales which migrate along the California coast each year engage in something called spy hopping. Observers believe the gray whales spy hop to check their location along the coastline, which might be termed line-of-sight navigation.

## Jumping for Air

Dolphins are the high jumpers of the sea. How and why they jump has been the subject of study for years. Scientists are still seeking more knowledge about this interesting behavior of dolphins.

To catapult themselves up and out of the water, they use the mighty thrust provided by their flukes; these are powered, in turn, by masses of lumbar muscles, connected with their tails by a series of tendons. Their rigid vertebral columns and these powerful muscles give the dolphins the strength and the speed needed to launch themselves.

Sometimes dolphins leap high into the air, reentering the water headfirst or belly flopping back. Sometimes they make flat jumps, arcing out of the water and splashing back in again.

There are two main speculations as to why dolphins behave in this manner. First, they

do it for the fun of it, jumping as a form of play. Second, they leap to get air, which they need because they are mammals. The second theory is explained by noting that simply rising to the surface for a breath of air without leaping clear of it causes a drag and turbulence, and jumping eliminates that additional drag. The first point seems just as valid in view of the astounding jumps for meager rewards that dolphins perform in captivity. Some captive dolphins have been noted jumping to a record of about 22 feet.

***Common dolphins.*** *This marvelous scene has been witnessed by many an ocean voyager. Dolphins leap out of the water seemingly for the sport of it—and to the delight of those watching.*

Orcas and pilot whales also submit to training for spectacular shows.

To leap to these heights above their pools, dolphins and whales require a takeoff depth of only about 20 feet, which suffices for them to gain the speed of 24 knots required for such performances.

## Skimming Above the Water on Underwater Wings

A hydrofoil boat is raised above the surface to escape the resistance of water. These craft start from a stationary position in the water exactly as conventional ships do. At low speeds they remain with their hulls in the water. But as they gather speed, submerged wings called hydrofoils provide lift and raise the hull clear of the water. These act in much the same manner as wings on airplanes, designed to take advantage of the imbalance of pressures above and below the foil, which at high speeds raises both foil and ship. A submerged propeller provides the ship's power.

The hydrofoils are generally one of two types: those that are fully submerged, and those that pierce the surface. Those fully submerged are very similar to airplane wings, fitted with ailerons that are con-

stantly adjusted by autopilot to maintain proper depth in all wave conditions. Surface-piercing foils are attached to the hull by struts that form a V or U and give the ship stability. This provides a very smooth ride in water that is not too choppy. Hydrofoils are used as passenger ferries in many of the world's waterways. They are employed by commuters in water-oriented cities like New York and Vancouver. They provide rapid crossings between Sicily and the Italian mainland, and Florida and the Bahamas.

***Popular ride.*** *One of Sea World/Atlantic Richfield's hydrofoils glides over Mission Bay near the marine life park in San Diego, California, at speeds as fast as 35 miles an hour.*

Despite their advantages over usual surface ships, hydrofoils are still of limited use. Their foils are subject to great stress at high speeds, and their size cannot be expanded without an undue increase in weight, which would require an enormous expansion of the power system. Hydrofoils are not yet economically feasible for general use.

## Riding a Cushion of Air

The Hovercraft, also known as the air-cushion vehicle or ACV, rides above the surface of the sea on a pad of compressed air provided by powerful fans. While the air keeps the craft above the surface of the water, huge aircraft propellers drive it at speeds far greater than those of conventional vessels. By remaining aloft, the ACV is free of resistance from the water, which reduces the efficiency of surface vessels and submarines. Hovercraft ferries cross the English Channel between Dover and Boulogne, Calais, or Dieppe on a regular schedule at speeds of up to 50 knots. The crossings of this normally rough body of water are either smoother than the ones made by other ferries, or impossible, if the sea is choppy. Hovercraft are very sensitive to the action of winds, and in storms they drift considerably. When a Hovercraft encounters waves higher than its normal flight altitude (which is proportional to the size of the craft), it must either settle

*Hovercraft.* *This man-made vehicle travels across water just above a cushion of air provided by the downward thrust of powerful fans.*

on the water or ride contour over the waves. Following the first alternative, the Hovercraft is subjected to the same water resistance as a surface vessel, and its speed is greatly reduced. Passengers in a Hovercraft flying contour over high waves are given an amusement park ride.

Hovercraft are coming into increasing use because they are able to function on ice, mud, and hard ground as well as over water. As their efficiency increases with size (doubling the size requires doubling the power, but the payload capacity is quadrupled), Hovercraft are expected to be the ocean cargo vessels of the future. With their great speed, they could outrun or circle major storm centers.

# Chapter X. Toward Higher Speeds

Man has used the oceans as a transportation route for thousands of years. For efficiency, it is unfortunate: the surface of the sea is the last place to navigate! Hulls are half submerged, half above the surface; and this creates monumental problems, among which is the necessity that the ship be a compromise between hydrodynamics, aerodynamics, and seaworthiness in storms. Practically no fast-moving animals have adapted to stay on the water to cruise. Surface ships sooner or later are bound to disappear—with the exception of yachts.

In underwater propulsion, to reach higher speeds and better efficiency, the challenge is to move through water without moving water, because moving water requires transferring some precious kinetic energy to the surrounding medium, which amounts to waste. We have seen that to achieve such swift journeys, fish, whales, or man-made submarines streamline their designs, avoid all protuberances, and develop very smooth skins or hulls. But this is only half the problem. Whether we consider a tunafish or a military torpedo, it will only move if thrust is applied, and the bigger the thrust, the higher the speed. To generate thrust, a power plant is needed, along with a power transducer system. Finally, some energy reserve must be stored to meet unpredictable demands.

Power plants are of three main generations: the cold power plant used by most cold-blooded animals, such as reptiles and fish; the warm power plant of birds or mammals, and the lukewarm power plant found in some rapid ocean fish like the yellow-finned tuna. In the cold-blooded system, there is practically no "heat overhead;" the energy needed by an immobile fish is negligible, and it could stay there for weeks without eating and suffer very little. Most of the food consumed by the conventional fish is used to grow, to reproduce, and to *move*. This seems to be an advantage over diving birds and mammals, like penguins and dolphins, which have to maintain a high central temperature even if they remain at rest, and cannot avoid substantial heat losses. But cold blood is seriously handicapped against warm blood if high speeds have to be maintained: high output power plants, whether man-made or natural, have a power-weight ratio that grows with the temperature of the "heat source;" no cold-blooded fish will ever compete with mammals in transoceanic marathons. The "lukewarm" yellow-finned tuna has developed a "counter current" heat exchanger system between venous and arterial blood wherein the heat is retained by transferring it from one system to the other.

The "power transducer" can be the jet funnel of the octopus, which develops considerable thrust, but has poor efficiency; the tail fin of a swordfish; or the "rotating fin" (the propeller) of a nuclear submarine. At least one-third of the energy is lost at the transducer level of the fast-swimming machine.

***Jaws*** *had a great influence on higher speeds. The most primitive fish, living over 300 million years ago, were probably very sluggish swimmers. With heavy armored bodies, they slowly meandered along muddy bottoms feeding on whatever organic debris was available. Their circular, jawless mouths were capable only of sucking. Speed was of no use. Eventually some fish developed a primitive movable mouth with jaws, allowing them to actively seek out other animals as a source of food. Those fish sought as prey then had to avoid capture by swimming; any stragglers were eaten, only the more rapid swimmers survived. Their success meant that only the quickest, most streamlined predators could survive in competing for food. If it were not for jaws, body shape, and propulsion, fish would bear little resemblance to what we see in these snappers.*

## Using Bursts of Speed

Bursts of speed are used by slow or sedentary creatures like the triggerfish or the octopus; by transoceanic cruisers like pilot whales; and by swift swimmers like the barracuda. To perform such "rushes," a creature must have the ability to generate a strong acceleration of its body, an instant tremendous thrust forward. The necessary energy must be readily available. It has to be stored, chemically, *inside* the muscles concerned; the energy conveyed by the blood from the central plant will come later to restore whatever was consumed locally. The translation of the energy into a large, short-duration force is best illustrated by the jet squirts of squids—they use such pulses constantly to prey on flyingfish at night or to escape porpoises, and they remain almost motionless between two darts. Another typical short-impulse transducer is the deep, long, soft tail of groupers. When they spring to attack an intruder in their territory, the first stroke of

their tail is so powerful that it creates a strong sonic boom due to cavitation. Such organic "explosions" are familiar to divers. Open-ocean fish that are well streamlined and powerfully equipped to cruise—like mackerel, tuna, and barracuda—sustain fairly high speeds as a routine, but are also capable of accelerating suddenly for short bursts of speed. These are needed mainly to catch a prey, to escape a more powerful predator, or for the incredible love dances of such strong fish as ocean jacks in the spawning season. This could not be achieved by those dark muscles that energize the tail at cruising speed because they receive just the amount of fuel that the central plant can deliver on a regular basis. Other muscles, light in color, are called for, and these have a built-in storage of energy.

***School of barracuda.*** *It is typical of barracuda to remain apparently motionless for a while and then dart quickly off without any apparent motive.*

## Shapes of the Fastest

It has been established earlier that all fast-swimming marine animals have the same basic streamlined body profile. To refine streamlining at high speed, the fastest fish have developed grooves or depressions where fins can be neatly tucked away so that nothing protrudes from the living projectile. The flow of water into the mouth and along the gills has also been eased by inner fairing in every detail. Another refinement is the development of laterally streamlined keels on the caudal peduncle: these add the finishing touch to the tuna's tail—a masterpiece of design. Broad and short, highly efficient when the maximum thrust is generated, the tail, in front of the fin, becomes horizontally compressed and tapered on both sides. Such a shape at the same time reinforces the tail, reduces the cross section of the inefficient part of the tail, and fairs it so that it knifes alternately with a minimum of water displaced. These caudal keels, shaped in "inverted streams," are found in such animals as marlin and killer whales, which have extremely different stories of evolution.

***A / Wahoo.*** *This member of the mackerel family has all the features of a typical fast fish.*

***B / Dolphin.*** *These members of the order Cetacea, like their larger relatives, whales and porpoises, are masters in the art of motion.*

***C / Pelagic shark.*** *Open-ocean sharks tend to be fast moving and possess the basic fusiform body shape most efficient for such a life-style.*

***D / Blue whale.*** *The largest creature on earth, this cetacean is not cumbersome but moves with grace and deceptive speed.*

***E / Swordfish.*** *A fine-pointed sword ahead of its body offers some degree of lateral streamlining to this fish.*

***F / Tuna.*** *The dorsal fin on the tuna fits into a slot in its back when the fish speeds off.*

***Mackerel*** *are slender, torpedo-shaped fish, which usually travel in fast-moving, well-disciplined schools. They are also fork-tailed.*

***Jacks*** *have laterally compressed bodies, forked tails with reinforcement at their base, and long, curved pectoral fins.*

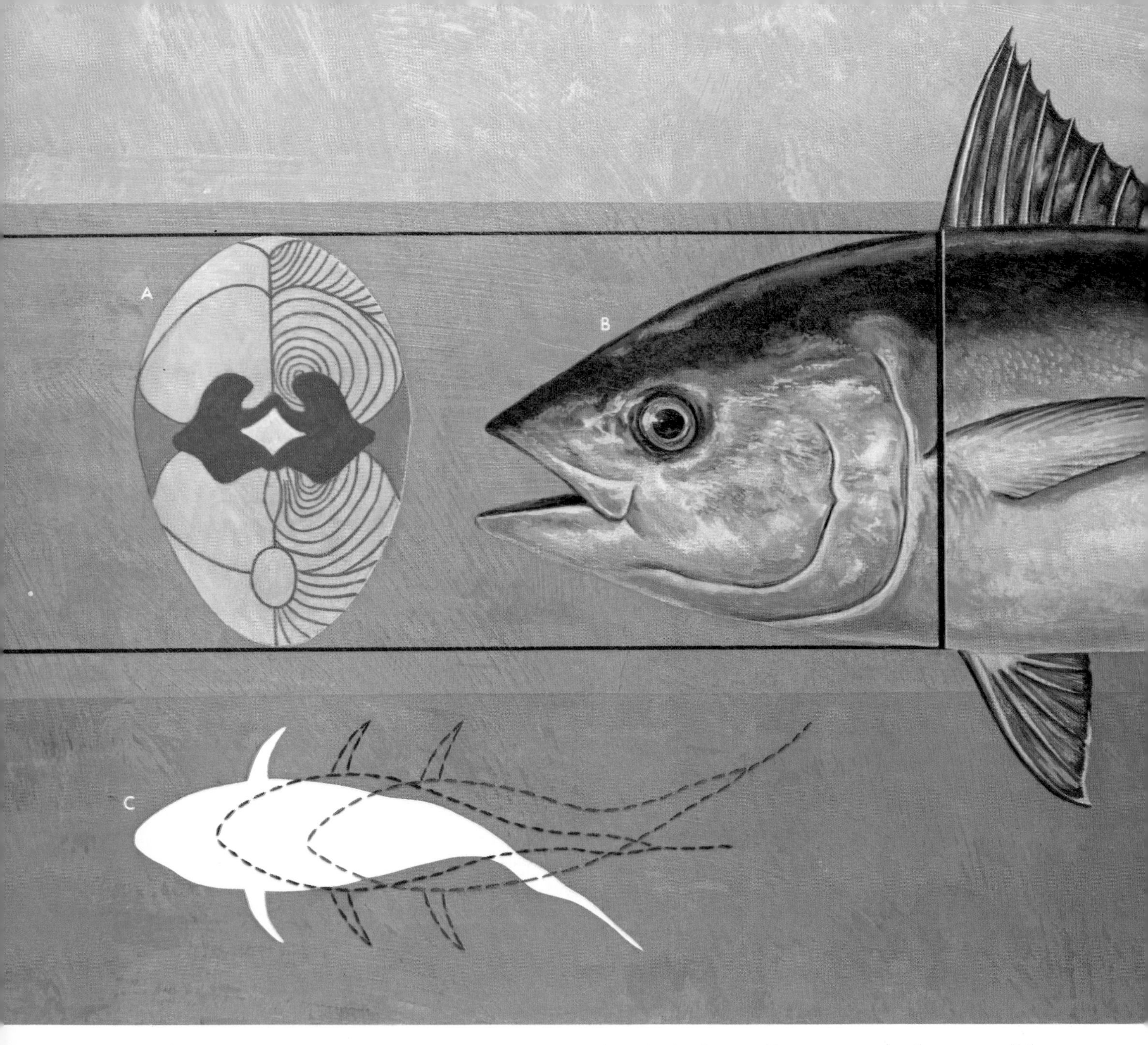


## Anatomy of a Tuna

The bluefin tuna is one of the fastest fish because of its design, the high "aspect ratio" of its tail fin, the reinforcement of its caudal peduncle, its powerful muscles, and the chemistry that supplies energy to fuel its fast motion. Laboratory measurements prove that the limiting factor is the amount of oxygen extracted from the water by the gills: per volume, water contains 25 times less oxygen than air. The mouth of the tuna is relatively small to create the least possible drag; the quantity of water it can pump in while swimming is proportional to speed. This explains why the tuna's metabolism is in equilibrium only at cruising speed. The tuna is sentenced to be a perpetual traveler. It has two different types of muscle tissue which provide a high gear and a low gear into which the tuna can shift.

***A / Cross section.*** *The tuna's powerful red tissue and white tissue are clearly shown. Red tissue probably provides the power for slow cruising speeds,*

*while the white muscles provide power for bursts of great speed. The red tissue derives its color from a bloodlike pigment that stores oxygen, providing greater energy. This means higher metabolism, which in turn means higher oxygen demand. To get the additional oxygen it needs, a tuna swims constantly, its mouth open to allow large quantities of water (and thus oxygen) to pass through its gills.*

***B / Side view.*** *The clusters of muscles, or myomeres, provide the thrust that drives the fish.*

***C / Powerful muscles.*** *These give a tremendous thrust to the tail and tail stalk, or hind portion, of the body. The front two-thirds of the fish remains rigid while the hind portion moves from side to side at a rate of about 15 beats per second when the fish is cruising. To achieve this motion, the tuna has two double joints—one at the base of the tail itself and the other at the base of the caudal peduncle.*

***D / Additional reinforcement of the skeleton.*** *A bony keel serves to stiffen the tail to prevent any flexing; it also aids the tendons in their function of moving the tail back and forth. By being stretched around the keel, they get increased leverage for pulling on the tail because the force is applied at a greater angle. In addition to the bony keel, there is a fleshy keel which provides the leading edge that smoothly separates the water as the tail beats from side to side.*

***Tarpon.*** *It is an endangered fish, relentlessly pursued by fishermen. The tarpon nurseries, in mangrove swamps of southern Florida, are destroyed.*

## The Endangered Silver King

The Atlantic tarpon is a shiny silverfish with extremely large scales and a deeply forked tail. It can reach more than eight feet and may weigh 250 pounds. It is a powerful migrator, related to the much smaller herring. It has a pouting appearance, due to its large protruding jaw. The high-metabolism silver king, hooked on a line, jumps high, pulls hard, and fights for a long time before agonizing at the mercy of the "big game" angler. As it is not edible, it is only caught for the pleasure of torturing, with endless refinements, one of the rarest and most beautiful creatures of the sea.

***School of smelt.*** *Only through effective swimming abilities do the defenses of schooling behavior and silver color become effective.*

## Slime for Speed

Most fish, like these smelt, have their scales—all along their bodies—covered with slime. Secreted through mucous glands in the fish's skin, the slime serves two functions. It acts as a lubricant to "grease" the fish's way through water with the least possible amount of friction. Being composed of long-chained molecules, slime helps stabilize water in laminar flow. Probably its most important function is that of sealing the fish and making it watertight. The skin of a fish is semipermeable, and without slime, the salt concentration in the fish would tend to equal that of the sea through osmosis.

***Dolphins*** *migrate across oceans at average speeds about double that of the cruising pace of the fastest cold-blooded fish.*

## Do Dolphins Defy the Laws of Hydrodynamics?

Performance of aquatic animals is usually estimated indirectly by calculating the drag forces on a fast-swimming fish or a marine mammal and then figuring the muscle power necessary to overcome such drag. From computations of this type emerged Gray's paradox, named for the famous scientist of animal locomotion. To explain the dolphin's ability to exceed speeds of 30 knots, this paradox pointed out either that the dolphin's muscles could produce ten times the power of terrestrial muscles (man can develop about .024 horsepower per pound of muscle) or that their bodies generated com-

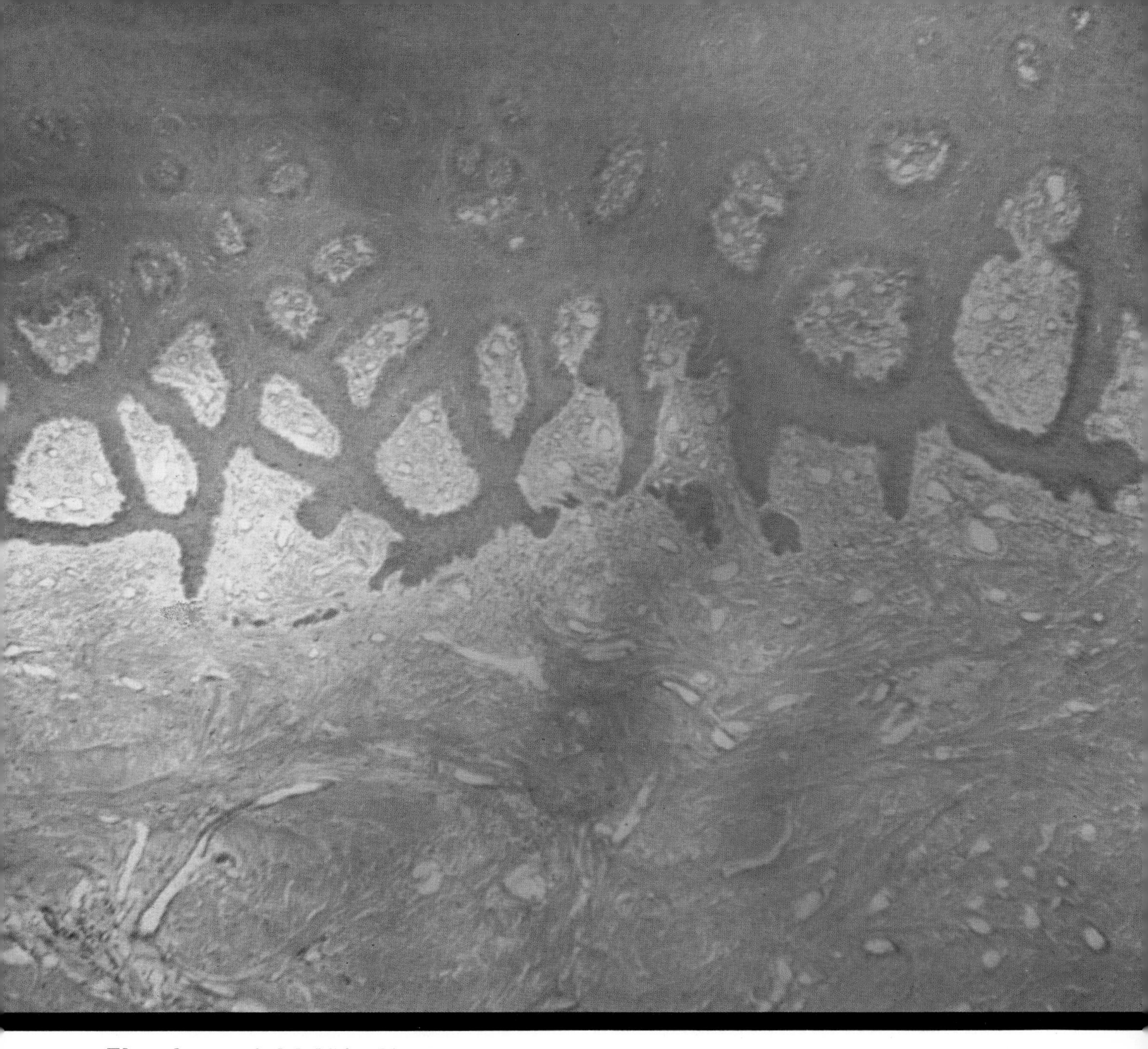

***Three layers of dolphin's skin.*** *Shown in this photomicograph at top is the epidermis, supported by layers below that can change the body contour.*

plete laminar flow. Pure laminar flow has never been observed at such speeds and body sizes as those of the dolphins.

How could Gray's paradox be solved? The figures for the dolphin's performance were not wrong; therefore the drag calculations had to be grossly on the high side. Recent studies have proved that marine mammals have a flexible and pressure-sensitive skin that dampens boundary-layer turbulence. The dolphin's skin has three layers: a thin, flexible outer layer; a thicker middle layer containing channels filled with a viscous substance; and a stiff, thick inner layer. When the boundary layer tends to thicken in eddies, the skin is depressed and it is pushed out if that layer thins out. This passive system requires no energy from the dolphin and dramatically increases its performance.

## Paddles

Sea lions and fur seals have retained four well-developed limbs from their terrestrial origin. They can still gallop on land faster than man can run. Underwater, they use their long, well-streamlined forelimbs as powerful paddles while the rest of their flexible bodies and their hindlimbs are mainly three-dimensional rudders. Antarctic penguins swim in a very similar fashion, using their wings as paddles. The sea turtle also paddles with its flattened front paws and leaves its hind paws trailing. Perhaps the most unique mammal propulsion system is that of the sea otter: its flat tail adds to the flattened hindlegs to constitute a triple rear-propulsion device.

***Fur seals.*** *Above, these mammals swim gracefully, using their forelimbs as wings and their hindlimbs as rudders.*

***Sea otter.*** *This mammal, at right, is well adapted to ocean life. Its tail is flat and has become a powerful propulsive aid.*

## Auxiliary Traveling Wave

Squids rival fish in their design for speed during brief runs, but they cannot compare with bonitos for long-range cruising. There is one field in which squids are unique: maneuverability. Capable of swinging their siphon nozzle almost 360°, they can aim the propulsive thrust of their jet and dart in any direction. But these members of the jet set have also adopted the traveling-wave system as an auxiliary power transducer for low speed. On each side of their bodies is a horizontal muscular fin that can be controlled to undulate in both directions, so that they can shift in low or high gear as well backward as forward!

***A / Moving forward.*** *Here the squid's siphon is pointing toward the rear of the animal, and when water is forced through it, the squid will dart forward.*

***B / Moving backward.*** *When the siphon is swiveled around toward the front of the squid and water is forced through it, the squid will shoot off toward the rear, its triangular fins acting as rudders.*

A
B

## The Fastest Underwater Vehicle: Nuclear Submarine

With its periscope, snorkel, radar and radio antennae retracted into its sail and all hatches battened down, the nuclear-powered submarine *Sculpin* moves through the sea, its decks awash. Only the sail protruding upward from the hull's rounded, tapered cylinder breaks the boat's clean lines. The hull is a deliberate and careful imitation of the general body streamlining of swift cruising, large animals—mainly whales. Buoyancy of the submarine can be adjusted by filling or emptying ballast tanks, which more or less replace the mammal's lungs or the fish's swim bladder. Static and dynamic stability are achieved, as in most conventional submarines, by ballast in the keel, diving planes, and rudders. The first great improvement was made at the power plant level, when the traditional, low-efficiency tandem "diesel on the surface" and "electric

*Nuclear submarine.* Designed for fast, efficient movement through the sea, it is no coincidence that the submarine bears strong resemblance to some of the ocean's mightiest whales.

undersea" was extended "diesel undersea" by the introduction of snorkels by the Germans in World War II. With the snorkel, the submarines were still breathing oxygen from the air, and remained in the sea mammal family. Then came the second, and maybe final revolution: the replacement of diesels and cumbersome electric batteries by a nuclear power plant. With unlimited range underwater, capable of producing oxygen from the sea, submarines became fish. Even with the mediocre efficiency of propellers as thrust transducers, the new power plant is theoretically capable of pushing the submarines at incredible speeds, but the limitation comes from the difficulty of keeping longitudinal stability under control at very high velocities. Gyroscopic, airplane-type controls of the diving planes make fine vertical steering possible.

# Chapter XI. Movement Shapes a Mode of Life

The form of an aquatic animal reflects its way of moving and its way of living. Thus, a slender, fusiform fish is likely to be a swift-swimming predator. Less streamlined creatures have adapted to a slower way of life. Similarly, a globular fish wallows, probably close to cover or otherwise protected from predators. The Portuguese man-of-war drifts on the surface with the wind. It has a sail, or float, which takes the wind the way a boat's sail does. Its trailing tentacles, catching what they can, slow it down and give it a better chance to sting and stun unwary creatures. The wind blows it anywhere just as it blows other drifters.

Crustaceans, crabs and lobsters, move in a cumbersome manner. Their armor doesn't permit flexibility of movement, so they lumber on all eight or ten legs, sometimes sideways, sometimes forward, their way of moving dictated quite strictly by their rigid coat. Consequently they seek food that is easy for them to get—slower-moving creatures or carrion. Or sometimes they dig. And burrowing animals display tools of their trade—limbs, heads, or appendages that aid in burrowing.

The skates and rays and other bottom dwellers—flounders, goosefish, sculpins, and catfish—are adapted for a life on the sea floor. With their flattened forms, they can stir up bottom sediments and let the mud or sand settle on their backs, hiding them effectively. And, as we have seen, their shape enables them to "fly" virtually undetected along the bottom. Others have settled on the sea floor for other reasons. They may have no swim bladder, for example, so they sink to the bottom and have adapted fins that help them plod along.

Where a fish lives has a great effect on its form. Or, conversely, perhaps the fish seeks a place where its attributes are best suited. The trout and salmon are fusiform fish with all the characteristics necessary for life in a swift-flowing stream or in ocean currents. They have pectoral fins close behind their heads. They have pelvic fins on their sides halfway back along the length of the body.

The articulations of the vertebral column of most fish have a much greater degree of freedom laterally than vertically, but it is exactly the contrary with marine mammals. Fish, therefore, can swing in a horizontal plane and have developed a vertical caudal fin; they are poorly equipped to change rapidly the level at which they swim, which is fortunate, because otherwise they could blow up their swim bladder. On the other hand, whales or porpoises can bend more in the vertical plane and have developed a horizontal tail; they constantly sound or breach, and this is a necessity for air breathing.

When a new creature is given a shape after a mutation, its propulsive capabilities and thus its way of life are a mere consequence of its anatomy.

***The open-sea shark** is built for speed. With a fusiform body which cleaves the water with minimum disturbance, this creature of the open sea can roam the world seeking smaller swift swimmers as food. In fact, because the pelagic shark has no swim bladder and its body is heavier than water, it would sink to the ocean floor if it didn't swim constantly and had no other equipment to counter the tendency to sink. But the large upper lobe of its tail fin provides some upward thrust; the pectorals, often long and broad, serve as planes that give it additional lift. The shark has yet another aid to float—an oversized liver. In some sharks the liver, containing lighter-than-water oil, reaches a weight of 1000 pounds. It is often 25 percent of the body weight and serves the animal as a buoyancy organ.*

## Awkward and Graceful

The manatee's awkward shape belies its grace in the water. Looking like a shapeless clumsy mass of flesh, the vegetarian manatee is a good swimmer. It has two short flattened forelimbs, just adequate to paddle slowly on the bottom or to help push back to the surface to breathe; but its enormous broad tail is incredibly powerful. It gives the manatee the ability to escape its very few predators.

Sea lions have streamlined, bullet-shaped bodies and an extremely flexible spine which enables them to make U-turns in less than a quarter of their length. Their front paws have become long winglike flippers and they literally "fly" underwater, in the same free style that swallows use in the air. Like their relatives the seals, sea lions have a fur coat: such a hairy surface was thought originally to increase the friction drag and it probably does so, but to a very small extent. However, it has recently been demonstrated that the fur padding acts as a boundary-layer stabilizer, reducing turbulence in very much the same way as the triple-decked skin of the dolphin does.

With their speed and maneuverability, sea lions spend a minimum of time feeding. They enjoy freedom of movement, and often tease other animals or play with them.

***Manatee.*** *This vegetarian mammal (above) eats the water hyacinth, a plant which, though harmless and beautiful, would otherwise flourish and clog the waterways around Florida.*

***Sea lion.*** *This engaging mammal (at right) is the picture of grace, power, and joy in the water.*

## Why Fish Face the Current

The ocean's waters are rarely still. They move in great currents that modify temperatures on a global scale. Fresh water flows from rivers that drain the continents. Eddies form on the fringes of major currents. Tides advance and recede as the moon circles the earth. Whatever causes water to flow, fish generally will face into the current. Mouths open and almost motionless, they only occasionally adjust a fin to maintain position. A fish facing a strong current can maintain itself in the same place with apparently less effort than would be required to swim at the same speed in still water. This can only be explained by comparing them with the large gliding seabirds, which are able to follow a ship for days without moving a feather. The fish take advantage of turbulence created by the current in water as the albatross benefits from eddies generated by the wind.

A fish facing into the current has a good

chance to capture a meal if that meal is moving directly toward the fish's mouth. Fish also extract oxygen from water that passes over their gills. Moving water, having been churned in air, is oxygen-rich. Of benefit too, in many cases, are temperature differentials that may enhance living conditions.

There is another advantage to facing the current: if a current were to hit the side of a fish, the animal would lose control and be carried along with the flow, perhaps into dangerous or intemperate locations. Facing away from the flow would similarly affect the fish's control of its speed and direction. Like more sessile animals, many fish let moving waters bring them the necessities of life, rather than pursue them.

***Schoolmasters.*** *These fish, which seem to be looking right at us, are a common species, growing to about 18 inches and weighing about three pounds. By facing upcurrent, as they are doing here, they don't have to work quite as hard for their food.*

## Becoming Sessile

Barnacles start out life as free-swimming creatures. Only later, as they develop into a near-adult stage, do they settle down. Then they attach themselves to some solid substrate, like rocks, whales' bodies, ships' bottoms, or wharf pilings. When they first hatch out of eggs, they resemble their crustacean relatives, the crabs, lobsters, and copepods.

At this nauplius stage they swim about freely. After several moults they change to the cypris stage. In this stage, the barnacle is in a bivalve shell with a muscle to hold the two valves or shells together. It also has developed a pair of compound eyes as well as a cement gland which it will use to glue itself to something solid. As an adult, the barnacle uses that gland to stick itself onto a solid substrate. It depends on the movement of the water around it or the movement through the water of the substrate it is attached to for food. Its bivalved covering has added plates that seal it in except at one end. From that end the legs protrude, kicking plankton into their mouths.

Some barnacles are minute in size. Others, such as those found on whales are rather large, reaching a size of six inches. A few species are parasitic on crabs. Some others are commensal with whales, turtles, or a few fish.

***Filtering feet.*** *Above, a barnacle has been removed from its protective armor, revealing its head region and highly modified feet, which are extended and retracted to filter.*

***Sessile colonies.*** *In its adult and more familiar form, the goose barnacle shown at right is found in colonies, securely attached, unmoving and almost immovable.*

## Angelfish

The tall narrow form of the queen angelfish governs its life-style. Fish of this family and others that have laterally compressed bodies maneuver easily. Their flat shapes enable these fish to slip into tall and narrow crevices in coral reefs to find protection from larger predators. Here they can also find the tiny morsels of food their small mouths can take in. Their flat bodies can fit into narrow horizontal crevices that they sometimes penetrate after leaning completely on their side to swim parallel to the bottom.

> **"These fish and others that have laterally compressed bodies are extremely maneuverable. They slip into tall and narrow crevices in coral reefs, finding protection from predators."**

## Flatfish

The flatfishes like this sole and its relatives, the flounders and halibuts, are in fact laterally compressed fish, although at a glance they give the impression of being dorsoventrally depressed. The swimming movement of a flatfish is much the same as that of other fish. It propels itself by bending its body from side to side in a traveling wave, moving horizontally rather than vertically. In its larval form it is no different from normal fish. It lives at the surface of the sea with one eye on each side of its head. Then one of its eyes begins to migrate to the other side and the shape of the fish gradually flattens. Finally it sinks to the bottom. Its eye having migrated completely, it lies over on one side and spends the rest of its life that way. Flatfish have a mode of life which is almost completely governed by their strange shape and position. The advantage they have is being able to hide on the sea bottom, covering themselves with sand.

## Sea Anemones

Sea anemones, relatives of the more mobile jellyfish, have become shaped to fit the sedentary life they live. Their form is stalklike, resembling a plant; in fact they are often mistaken for plants. They can move if they must, but usually they remain in one place for extended periods of time. Generally they live a quiet and sessile existence, taking food from water that passes by them. When they do move, it is with great awkwardness. To move, some must release the hold their pedal disc has on the substrate; they then fall over on one side, twist around, and reattach their pedal disc in a new location. Others merely inch along by snaillike movements of their pedal discs. Despite this inability to move along freely and easily, the sea anemone is very well protected. Its stinging tentacles, which stun its prey, can also be retracted when the animal is not feeding. Delicate, decorative looks are more than compensated for.

## Little Fish on Reef

Why do so many little fish live around reefs? It may be because it is an ideal place for small fish. On a reef, whether it is made up of coral or of geological rock, there are many tiny cracks, slots, and crevices that serve as cover for small fish but not for large ones. Vast numbers of microscopic plants and animals around these reefs serve as food for the small reef fishes. Consequently their mode of life does not require that they be great swimmers; to avoid capture, they must only be able to dart into a nearby hole. For instance, in this tropical reef, the clownfish remains close to the anemone's stinging tentacles. In this marvelous relationship the clownfish may lure prey to the anemone, and the anemone protects the clownfish from predators. Dazzling color and variety abound in and around coral reefs, and many of the creatures living in these communities are camouflaged and assisted by their surroundings.

# Moving Elsewhere

I plunged into the sea and I forgot my load
I reached for a freedom that stunned and froze my heart

Freedom in three dimensions, cold and thick
And liquid embraces that stalled my skin

A dash of fluttering broth
A pulsing crystal froth

Spirit of the sixth sense
Speed spreading insolence

The cavalcade of shrimps
Flesh belt's extravagances
Rainbows of darting squids
Anonymous dances

And vertical gold leaves
Heading into the stream

Instinctive snouts gaping for oxygen and prey
Throngs of rhythmical fins beating mindless pathways

Then dolphins, seals, and whales stole fire
from the sun
To playfully invade the automatic sea
In their burning entrails our
powerful cousins
Presented their mother with songs, luxury,
and fun

But since I discovered the key to fluid dreams
The warmth of the whale's blood simmers deep in my heart.

While diving birds regret their weight
And leaping fish plane in the wind
Moving elsewhere
Elsewhere for hope.

# Index

A

Adélie penguins, 82
Air-cushion vehicle (ACV), 110-111
Albatross, 134
American eels, 22
Anemones, 88, 140, 141
Angelfish, 22, 138
Angel sharks, 16, 17
Aqualungs, 76
Archimedes, 12, 13

B

Baleen whales, 26
Barnacles, 88, 136
  goose, 136
Barracudas, 22, 114, 115
Bass
  sea, 23, 46, 47
  striped, 25
Batfish, 67
*Bathypterois,* 75
Big-eyes, 23
Billfish, 23, 100-101
Birds
  defiance of gravity by, 14-15
  water, 76, 94, 112, 134
Bivalves, 62
Blackfish, 41
Bluefin tunas, 118-119
Bluefish, 25
Blue whales, 10, 94, 116
Boobies, 15
Brittle stars, 75
Buoyancy, 12-13, 47
Butterflyfish, 22

C

*Calypso,* 60
Cardinalfish, 15, 23
Catfish, 130
Clams, 62, 68
Clownfish, 141
Codfish, 75
Coelacanth, 62
Commensalism, 91, 136
Conch, 62
Conger eels, 22
Cormorants, 15, 82
Cornetfish, 24
Cowfish, 22, 30, 32, 33, 94
Crabs, 64, 130, 136
  hermit, 88
Crayfish, 64
Crinoids, 75

D

Dinoflagellates, 76
Diving saucer, 54, 60
Dolphins, 8-9, 20, 26, 27, 28, 36, 37, 76, 88, 92-93, 106-107, 112, 116, 122-123, 132
Dorados, 96
Ducks, 94

E

Eels, 30, 38
  American, 22
  conger, 22
  moray, 24, 50
Einstein, Albert, 9
Electric rays, 17
Evolution, 8, 26, 80

F

Filefish, 22
Fins, role of, 38-53
Fish lice, 88, 90
Flatfish, 139
Flatworms, 88
Flounders, 130, 139
Flyingfish, 96-97
Fugu, 51
Fur seals, 62, 124

G

Game fish, 100
Geese, 94
Goose barnacles, 136
Goosefish, 25, 130
Gray's paradox, 122-123
Gray whales, 10, 27, 105
Gribbles, 90
Groupers, 40, 47, 91, 114-115
Grunts, 16
Guitarfish, 22, 25
Gulls, sea, 14

H

Haddock, 41
Hagfish, 30
Halibuts, 139
Hardtails, 25
Hermit crabs, 88
Hovercraft, 110-111
Humpback whales, 27, 105
Hydrofoil boats, 108-109

I

Iguana, marine, 80
Isopods, 90

J

Jacks, 38, 40, 115, 117
Jaws, 112
Jellyfish, 54

K

Killer whales, 36-37, 107

L

Lampreys, 30, 88
Lice
  fish, 88, 90
  whale, 88
Lizardfish, 23
Lobsters, 64, 65, 130

M

Mackerels, 25, 27, 40, 115, 117
Mako, 25
Manatees, 132
Man-of-war, Portuguese, 130
Manta rays, 17, 83, 88
Marine iguana, 80
Marlins, 23, 38, 100
Medusas, 54
Molluscs, 69
Moorish idols, 42
Moray eels, 24, 50
Mudskippers, 62
Mullet, 102
Mussels, 62
Mute swans, 94

N

Narwhal, 27
Nudibranchs, 68, 69

O

Octopus, 54, 58-59, 62, 112, 114
Orcas, 36-37, 107
Ostraciiform movement, 33
Oysters, 62

P

Pacific salmon, 98-99
Parasites, 88, 90, 91
Parrotfish, 25
Pelagic sharks, 116, 130
Penguins, 76, 82, 112
  Adélie, 82
Petrels, 15
Phytoplankton, 88
Pilotfish, 88
Pilot whales, 107, 114
Pipefish, 24
Porcupinefish, 22, 33, 51
Porpoises, 26, 27, 78, 130
Portuguese man-of-war, 130
Protozoans, 62
Pufferfish, 33, 49, 51

R

Rays, 16, 17, 25, 38, 130
  electric, 17
  manta, 17, 83, 88
  sting, 25
Reef sharks, 44
Remoras, 91
Requiem sharks, 44
Ribbonfish, 30
Rockfish, 38
Roundworms, 88

S

Sailfish, 23, 100
Salmon, 35, 130
  Pacific, 98-99
Salps, 54, 57
Sardines, 94
Sawfish, 25
Scallops, 56, 62
Schoolmasters, 135
*Sculpin* (submarine), 128-129
Sculpins, 130
Sea anemones, 88, 140, 141
Sea bass, 23, 46, 47
Sea cucumbers, 68, 69, 72, 73
Sea gulls, 14
Seahorses, 22, 38, 53

Sea lions, 9, 52, 62, 103, 124, 132
Seals, 76, 132
fur, 62, 124
Sea otter, 124, 125
Sea robins, 66
Sea snakes, 30, 80
Sea stars, 56, 69, 70-71, 75
Sea turtles, 84, 91, 124
Sea urchins, 72-73, 75
Sharks, 17, 28, 75, 88, 91, 94, 130
angel, 16, 47
*Carcharhinus,* 42
pelagic, 116, 130
reef, 44
requiem, 44
whale, 91
Shrimps, 64, 65, 75
Skates, 17, 22, 25, 38, 130
Smelt, 121
Snails, 62, 68
Snappers, 112
Soldierfish, 4, 23
Sole, 139
Spadefish, 22
Sperm whales, 27, 105
Squids, 8, 10, 54, 114, 126-127
Squirrelfish, 23
Starfish, 70-71
Stingrays, 25
Striped bass, 25
Sturgeons, 41
Submarines, 76
nuclear, 128-129
Suckerfish, 91
Sunfish, ocean, 49, 52
Surgeonfish, 48
Swans, 18
mute, 94
Swim bladders, 16, 47, 130
Swordfish, 23, 100, 112, 116

T

Tangs, yellowtail, 48
Tapeworms, 88
Tarpons, 120
Terns, 15
Threadworms, 88
Triggerfish, 49, 114
Trout, 130
Trumpetfish, 24
Trunkfish, 32-33
Tunas, 8, 17, 18, 25, 28, 38, 40, 41, 112, 115, 116
anatomy of, 118-119
bluefin, 118-119
yellow-finned, 112
Turtles, 76, 88, 136
sea, 84, 91, 124

V

Velella, 86

W

Wahoos, 116
Walruses, 62, 81
Whale lice, 88
Whales, 26, 27, 36, 88, 91, 104-105, 112, 130, 136
baleen, 26
blue, 10, 94, 116
gray, 10, 27, 105
humpback, 27, 105
killer, 36-37, 107
pilot, 107, 114
sperm, 27, 105
Whale sharks, 91
Whelks, 68
Worms, 68-69, 75, 88
Wrasses, 23, 90

Y

Yellow-finned tunas, 112
Yellowtail tangs, 48

## ILLUSTRATIONS AND CHARTS:

Sy and Dorothea Barlowe—34-35, 118-119; Howard Koslow—21, 22, 23, 24, 25, 26-27, 33, 40, 41, 42, 52, 53, 62, 79, 96-97, 99, 116-117, 127.

## PHOTO CREDITS:

Ken Balcomb—36-37, 104-105; John Boland—39; Ben Cropp—31, 91; David Doubilet—24 (top), 134-135; F. Ferro, M. Grimoldi, Rome—139; Freelance Photographers Guild: Alpha—12, Shayne Anderson—69, Ron Church—55, P. Colin—117 (top left), Robert Evans, Western Marine Labs—137, FPG—101, Bob Gladden—66, 67, Tom Myers—68, 110-111, Chuck Nicklin—73, 89, 92-93, 133, R. Panouska—78-79, Siebe Rekker—95, John Stormont—53, 136, A. B. Trail—63, 90, 113, John Zimmerman—85, 98; George Green—141; Edmund Hobson—24 (bottom), 50; Holger Knudsen, Marine Biological Laboratory, Helsinger, Denmark—64; Bill McDonald—17; Jack McKenney, Skin Diver Magazine—47, 52, 57; Maltini-Solaini, M. Grimoldi, Rome—70-71 (top), 100; Richard Murphy—121; Naval Undersea Center, San Diego, Calif.—45, William E. Evans—106-107, John C. Sweeney—123; Carl Roessler—32, 46, 126-127; Sea Library: Ben Cropp—83; Sea World, Inc., San Diego, Calif.—108-109; Tom Stack & Associates: Ron Church—74-75, 117 (top right), 125, LeRoy French—65, M. J. Gilson—28, Ben Goldstein—82 (right), Richard F. Gunter—14, Jack McKenney—138, Tom Stack—19, William Stephens—86, 97 (top), 120; Submarine Flotilla One, Public Affairs Office, San Diego, Calif.—128-129; Paul Tzimoulis—48, 72, 77, 84.